AI and Sin

How Today's Technology Motivates Evil

Christopher M. Reilly, Th.D.

En Route Books and Media, LLC
Saint Louis, MO

⊕ENROUTE
Make the time

En Route Books and Media, LLC

5705 Rhodes Avenue

St. Louis, MO 63109

Contact us at **contactus@enroutebooksandmedia.com**

Cover Credit: Christopher M. Reilly with Google Gemini

Copyright 2025 Christopher M. Reilly

ISBN-13: 979-8-88870-345-8

Library of Congress Control Number: 2025933632

Table of Contents

Chapter 1

Introduction to the Argument

It would take a great deal of mental effort to ignore the flood of news about artificial intelligence technology (AI) in recent years. Few technological developments in history have generated such a level of popular excitement, futuristic imagination, and fear. Ray Kurzweil, computer scientist and author, has eagerly claimed that "we are going to expand intelligence a millionfold by 2045 and it is going to deepen our awareness and consciousness."[1] Eric Schmidt, the former CEO of Google, has publicly worried that AI systems will somehow gain independent control over dangerous and ultimately harmful means of destruction, declaring that "after Nagasaki and Hiroshima, it took 18 years to establish treaties banning nuclear tests. We don't have that kind of time today."[2] With similar concerns, the managing director of the World Economic Forum declared in 2023 that "93 percent of cyber leaders, and 86 percent of cyber business leaders, believe that the geopolitical instability makes a catastrophic cyber event likely in the next two years."[3] This sense of an impending, world-changing consequence is why sixteen leaders of world religions

---

[1] Zoë Corbyn, "AI Scientist Ray Kurzweil: 'We Are Going to Expand Intelligence a Millionfold by 2045,'" *The Guardian* (June 29, 2024), https://www.theguardian.com/technology/article/2024/jun/29/ray-kurzweil-google-ai-the-singularity-is-nearer.

[2] Brian Foster, "The Former CEO of Google Warns of a Global Catastrophe Caused by AI in Five Years," *Glass Almanac* (November 12, 2024), https://glassalmanac.com/the-former-ceo-of-google-warns-of-a-global-catastrophe-caused-by-ai-in-five-years/.

[3] Jeremy Jurgens, press conference, https://www.weforum.org/meetings/world-economic-forum-annual-meeting-2023/sessions/press-conference-global-cybersecurity-outlook-2023/.

gathered in 2024 in Hiroshima, the site of a nuclear bomb explosion that ended World War II, to sign the Catholic Church's document named *The Rome Call for AI Ethics*.[4] Elon Musk, the multibillionaire and head of the federal Department of Government Efficiency who is also a consistent proponent of AI, has admitted that "AI is a significant existential threat."[5]

Although I don't subscribe to either the breathless or the darker hyperbole about AI, the argument that I will make in this book may appear to be just as controversial and alarming. Readers – especially thoughtful Christians – will nevertheless find that it is derived from quite sober, thorough analysis, theological scholarship, and a plethora of evidence. What I am suggesting, in a nutshell, is that the presence, development, and use of AI technology generally and significantly motivates sinful action.

AI includes the machines, programs, and systems that have a relatively substantial capacity to emulate the powers of human thought in carrying out various tasks. Such efforts require learning, predicting, planning, communicating, performing semi-autonomously, and generating unique actions or outputs like text and images. Unlike other technologies, AI is not defined by the nearly endless array of uses to which it may be applied, and not entirely by the methods and machine structures that support it, but primarily by the extent to which such machines or devices have the capacity to generate output (text, images, calculations, inferences, physical actions, etc.) similarly to intelligent human beings. An AI system may therefore approximate or exceed human persons in its accomplishment of certain tasks, and the character of its "intelligence"

---

[4] RenAIssance Foundation, "AI Ethics for Peace – Hiroshima, July 9, 2024," https://www.romecall.org/ai-ethics-for-peace-hiroshima-july-9th-2024/.

[5] Marco Quiroz-Gutierrez, "Elon Musk Says There's a 10% to 20% Chance that AI 'Goes Bad,' Even While He Raises Billions for His Own Startup xAI," *Fortune* (October 30, 2024), https://fortune.com/2024/10/30/elon-musk-ai-could-go-bad-existential-threat-xai-fundraising/.

resembles the thinking capacities of human beings in some meaningful way. Examples of AI applications include chatbots that respond with intelligible text to users' queries or statements, sometimes assuming a persona in ongoing "conversations"; digital assistants that accomplish computer tasks as instructed by the users' voice and text prompts; interactive robots that may sometimes physically resemble humans; self-driving vehicles; eyeglasses that provide aural information, augmented vision, and guidance about the wearer's environment; personalized recommendations systems that help to customize user experiences from selection of music to social media newsfeeds; and predictive analytics that enables researchers and corporations to identify future outcomes and behaviors with remarkable accuracy.

Despite the amazing new abilities that human persons have acquired with the assistance of AI systems and machines, the amount of public ethical concern about the development and use of AI is striking. It is directly related to the wide-ranging and varied power that is expected to be wielded by persons, corporations, governments, and criminals who master its applications. For example, according to Gallup, more than three-quarters of American adults surveyed have little trust that corporations will use AI responsibly.[6] Ethical anxiety is widespread even though most emphasis in AI development is currently on rather benign consumer, healthcare, and business applications (with the exception of AI-enabled military weapons). This perspective is represented well by one senior scientist's comments: "My view is that technology can be used in so many different ways, and rarely do we understand the future implications of what we have made, even if we have a clear sense of intended uses. A

---

[6] Julie Ray and Gallup, "Americans Express Real Concerns About Artificial Intelligence" (August 27, 2024), https://news.gallup.com/poll/648953/americans-express-real-concerns-artificial-intelligence.aspx.

creative person can and will come along and use it in ways we never imagined."[7]

In short, many in our society are worried about the harmful effects of bad use of AI. Such worries derive from a view of AI as essentially a tool that people with various intentions can choose to utilize for either good or bad applications. This is far, however, from recognizing that AI has a more general impact of engagement in sin. After all, how can AI be sinful or evil if it is just a tool, a morally neutral means for accomplishing various ends? Isn't the user of a tool morally responsible, not the tool itself?

Regarding moral evil, which is the active, willful engagement in evil behavior, I wholeheartedly agree that moral responsibility can only be attributed to human beings who have the capacity for reason and free choice; this is a foundational principle of Christian morality. Let's be clear that neither this book's title nor any of my arguments will indicate that AI itself is inherently sinful or evil – only that its development, use, and proliferation throughout human activities motivate people, far too often and systematically, to engage in sin. We therefore might go so far as to describe AI as "an evil" since it is a real thing (or structure of things) that causes extensive harm. Calling something an evil because it tends to generate bad effects or consequences is not the same, however, as saying that it is evil in itself. Only personal forces are morally evil – or, to be even more precise, they may intend or enact evil behaviors.

Indeed, throughout human history, our technological tools, machines, and systems have usually been neutral means for accomplishing whatever ends we set our minds to. We need to be clear, however, just

---

[7] Natalie Klym, "The Technologists are Not in Control: What the Internet Experience Can Teach us about AI Ethics and Responsibility," in *The State of AI Ethics Report*, Volume 6, Montreal AI Ethics Institute (January 2022), 30, https://montrealethics.ai/wp-content/uploads/2022/01/State-of-AI-Ethics-Report-Volume-6-February-2022.pdf.

what we mean by "neutral." Tools do, after all, influence, guide, or limit human behavior. The famous philosopher of technology Albert Borgmann expressed this reality in an interview with *The Christian Century* in 2003:

> A crucial feature of a technological device is that it makes something available to us in a comfortable way. You don't have to work for it. It's there at our beck and call.
>
> That transformation is much more profound than we realize. If two or three hours of television a day come into our lives, then something else has to go out. And what has gone out? Telling stories, reading, going to the theater, socializing with friends, just taking a walk to see what's up in the neighborhood.
>
> ... It's an inducement, and it's so strong that for the most part people find themselves unable to refuse it. To proclaim it to be a neutral tool flies in the face of how people behave.[8]

Sometimes, the moral, social, and personal relevance of a technology may be difficult to see. In a simple example, the decisions made by the designer of a jackhammer and the physical operations of the machine will ultimately restrict my use of the tool to specific purposes like breaking up a sidewalk so it can be repaired. A jackhammer is nevertheless just a jackhammer, with little inherent *moral* relevance, unless I somehow decide to do something immoral with it that hurts another person. For most technologies and tools, it is hard to even imagine considering an immoral use

---

[8] David Wood, "Albert Borgmann on Taming Technology: An Interview," *The Christian Century* (August 23, 2003), 22-25.

for them (who would rationally choose to wield a cumbersome jackhammer as an instrument of violence?). Other technologies like military weapons or financial trading systems through the electronic stock markets enable many potentially immoral purposes, yet it nevertheless seems that it is not the technology that is immoral but the person who chooses to use the technology badly.

Even the Catholic magisterium has had to wrestle with this apparent conundrum in which advanced technology has both good and bad effects and does not have a determining control over human behavior, yet it is often associated with a way of being that steers us from our true path in developing a relationship with God. This is expressed in the Catholic Church's document *Donum Vitae*:

> It would on the one hand be illusory to claim that scientific research and its applications are morally neutral; on the other hand one cannot derive criteria for guidance from mere technical efficiency, from research's possible usefulness to some at the expense of others, or, worse still, from prevailing ideologies. Thus science and technology require, for their own intrinsic meaning, an unconditional respect for the fundamental criteria of the moral law: that is to say, they must be at the service of the human person, of his inalienable rights and his true and integral good according to the design and will of God.[9]

---

[9] Congregation for the Doctrine of the Faith, *Instruction "Donum Vitae" on Respect for Human Life in Its Origin and on the Dignity of Procreation: Replies to Certain Questions of the Day* (February 22, 1987), Introduction, 2.

Even though our technological artifacts and systems are "neutral" in the sense that they do not have a moral will of their own or a completely decisive influence on human behavior, we also need to recognize that, beyond the physical and structural influences on our behavior, there are many ways that a technology may influence our actions in a good or evil direction. The writer George Grant expresses this problem with "neutrality":

> When we represent technology to ourselves as an array of neutral instruments, invented by human beings and under human control, we are expressing a kind of common sense, but it is a common sense from within the very technology we are attempting to represent. The novelness of our novelties is being minimized. We are led to forget that modern destiny permeates our representations of the world and ourselves. The coming to be of technology has required changes in what we think is good, what we think good is, how we conceive sanity and madness, justice and injustice, rationality and irrationality, beauty and ugliness.[10]

The existence of nuclear weapons is a classic example of how the mere presence of a certain technology and the capabilities it endows us with can tempt people to acquire, threaten to use, or actually employ the technology to great harm of others. Without the technology, the temptation would likely not arise. Furthermore, if we did not have the specific, technical character of nuclear weapons as mass-destructive and presenting high scientific and financial barriers that give certain countries

---

[10] George Grant, "Thinking about Technology," *Technology and Justice* (Notre Dame, Indiana: University of Notre Dame Press, 1986), 32.

tremendous advantages over others, those countries would likely not imagine that they can feasibly threaten other countries with annihilation. Other countries and non-governmental actors would not feel exposed and seek out possession of nuclear weapons as instruments of self-defense or terror.

Consider also the expansion of pornographic technology through the internet and AI applications, leading to fake images and videos (deepfakes) that seem to portray real individuals without clothing and the many instances of malicious sharing of those images among the public. Without the internet, it would not be possible to share the images; without AI, it would not be possible to create them; and without the specific applications that are designed to enable creation of such images, few people would even think about such a goal. At some point of specification, at least some of these technologies are no longer neutral in a certain sense, because they have a negative motivational effect on persons. Human beings have free will, but we are all too easily influenced by our advanced technologies.

A third example is the variety of currently popular smartphone applications that enable and encourage regular reading of the Bible. Despite the very good intentions of such apps, some people have recognized that "Bible verse of the day" features have a strong tendency to favor only the brief and immediately attractive passages in the Bible while ignoring the more difficult, obscure, or simply longer passages. One article suggests that the apps are programmed to emphasize the most-shared verses that are frequently the most soothing and emotionally therapeutic.[11] This creates a "therapeutic filter bubble" in which the same, "feel-good" verses are

---

[11] Pete Phillips, "Why Your Bible App's 'Verse of the Day' Feature Could Be Skewing Your View of God," *Premier Christianity* (October 2, 2018), https://www.premierchristianity.com/home/why-your-bible-apps-verse-of-the-day-feature-could-be-skewing-your-view-of-god/3511.article.

repeatedly displayed. Christians, however, would be well-served by assistance and encouragement in reading and contemplating the Bible passages that require reflection, study, and an investment of time. Another article warns against a superstitious reliance on fate to determine which Bible passages one reads:

> Ancient Romans developed the practice [of *Sortes Virgilianae*], which involves opening a volume of Virgil to a random page and taking it as a personalized oracle.
>
> Performing a *Sortes Biblicae* by algorithmic rather than random means may be an even graver sin. The one who does so expecting a message from God conflates *vox populi* with *vox Dei*, mistaking a proprietary computer program that runs on user inputs for the inscrutable will of God.[12]

Whether or not the Bible apps are providing a valuable service to Christians, it is clear that the effect of the technology as an influence on persons' spiritual life is not precisely neutral nor insignificant.

In this book, I'll argue that there is something particularly concerning that is going on with AI technology. The risk of doing evil actions is clearly enhanced with AI. I'm not the only one who is concerned; consider the following statement by Pope Francis:

> "Intelligent" machines may perform the tasks assigned to them with ever greater efficiency, but the purpose and the meaning of their operations will continue to be determined or enabled by human beings possessed of

---

[12] Grayson Quay, "Algorithmic Spirituality," *First Things* blog (June 7, 2024), https://www.firstthings.com/web-exclusives/2024/06/algorithmicspirituality.

their own universe of values. There is a risk that the criteria behind certain decisions will become less clear, responsibility for those decisions concealed, and producers enabled to evade their obligation to act for the benefit of the community. In some sense, this is favoured by the technocratic system, which allies the economy with technology and privileges the criterion of efficiency, tending to ignore anything unrelated to its immediate interests.[13]

Francis is concerned that we will be tempted to allow and even enable an unlimited expansion of technological development through AI. A crucial part of morality, however, is setting limits on our desires and our striving for some kind of imagined (and usually distorted) perfection.

Human beings are, by definition, mortal; by proposing to overcome every limit through technology, in an obsessive desire to control everything, we risk losing control over ourselves; in the quest for an absolute freedom, we risk falling into the spiral of a "technological dictatorship". Recognizing and accepting our limits as creatures is an indispensable condition for reaching, or better, welcoming fulfilment as a gift. In the ideological context of a technocratic paradigm inspired by a Promethean presumption of self-sufficiency, inequalities could grow out of proportion, knowledge and wealth accumulate in the

---

[13] Francis, Message of the Holy Father for the 57th World Day of Peace on January 1, 2024 (December 14, 2023), 4, https://press.vatican.va/content/salas-tampa/en/bollettino/pubblico/2023/12/14/231214a.html.

hands of a few, and grave risks ensue for democratic societies and peaceful coexistence.[14]

But there is even more to the concern over AI. Such new technologies
directly challenge us in the moral sphere of life. For example, in the very
definition of AI presented above, there are some implications for real
moral issues, such as the designation of AI as focused on completing
"tasks." Although some researchers are straining to develop AI-based machines, especially robots, that resemble human beings in every way, the
usual purpose and very existence of AI is generally intended to assist humans in completing tasks. But why are we so concerned with devoting
trillions of dollars and entire economic sectors to the completion of such
tasks? Which tasks should we focus on? Who will benefit from the AI
assistance, and how will others be neglected? Shouldn't we instead be fo-
cusing our energies on charitable care for our disadvantaged neighbors,
personal growth in virtue and communal growth in solidarity, and contemplative worship of our God? These are all moral questions. Of great
moral concern is therefore the fact that the answers to most of these questions are effectively decided – at least to a considerable extent – by researchers, designers, coders, corporations, and others in the formation of
AI technology; the restriction of moral autonomy for the rest of us is a
moral problem in itself.

A very significant moral aspect of AI is the way that it affects our understanding of ourselves, both as individual selves who communicate and
live in a hyper-technological culture and as human beings. Pope Francis
has observed that "[AI] developments such as machine learning or deep
learning, raise questions that transcend the realms of technology and en-
gineering, and have to do with the deeper understanding of the meaning
of human life, the construction of knowledge, and the capacity of the

---

[14] Francis, World Day of Peace 2024, 4.

mind to attain truth."[15] When we compare AI capabilities to our own intelligence and even our own agency as independent, responsible persons, we inevitably encounter existential questions. What does it mean to be a person? Or to be truly intelligent? Is intelligence the same as wisdom? Where does the truth that guides and orients us come from – the factual data and calculations that drive AI machines, the social relations and negotiated meanings of our communities, our intuition and imagination, the natural law, the transcendent reason of God, or some combination of all of these? This is why the Church has applied such an extraordinary and intense level of attention to this very new technology: it challenges our understandings of who we are and it exposes the very narrow, misguided definitions of humanity that are typically assumed by the fields of experimental science, biology, and neurology.

In fact, the labeling of AI technology as "neutral" is closely associated with a reduced notion of human nature as being primarily intellectual, merely a sophisticated processor of information (in the form of data) similar to a computer. Michael Dominic Taylor explains that, "in the same way that money is presumed to be neutral, so too is technology, precisely due to the logical dualism it enforces between the intellect and will, and thus between truth (now the pragmatic notion of 'it works') and goodness (determined subjectively)."[16] This dualism between intellect and will is a philosophical idea that builds on René Descartes and runs through Immanuel Kant and his followers, which gives priority to the mind – the power of logical or calculative reason – and treats the human will as a purely independent engine of personal choice, governed only by a tentative conscience but unaffected by the reality of the human person as a created being in a created world. In other words, the morality of choices

---

[15] Francis, World Day of Peace, 2024.

[16] Michael Dominic Taylor, "'Riveted with Faith unto Your Flesh': Technology's Flight from Actuality and the Word Made Flesh," *Communio* 49 (2022), 539.

is not guided by any natural law, itself reflective of the Divine law, but it seems to be a radically free decision of human persons who "rationally" determine what is good. Under such a limited form of reason and the emphasis on intellectual processing of information, this understanding of the human person leads to a utilitarian focus that celebrates technological progress in measuring, controlling, and extracting material prosperity from all aspects of life and our environment. Taylor explains further:

> This account, while apparently true when considering morality in a vacuum, discards the ontological considerations that undergird ethics to begin with: namely, the normativity of human nature. In other words, to take a neutral stance toward technology generally is to claim that reality and varying levels of abstraction from reality are equally valid paths to human fulfillment. Moreover, to take a neutral stance toward technology is to take an agnostic stance toward human nature itself, reserving for the will alone the determination of what is good.[17]

The understanding of nature in the digital age as fundamentally comprised of "information" alters the way we construct and infer meaning in our use of technological artifacts. Nadia Delicata refers to the famous theory of Catholic philosopher Marshall McLuhan that all tools are essentially extensions of the persons who use them, in space and time (think of wielding a hammer) but also in the way that they express the intentions of the user, and they form part of an environment that influences persons'

---

[17] Michael Dominic Taylor, "'Riveted with Faith unto Your Flesh,'" 539.

actions and interactions (think of how Facebook algorithms favor certain posts or connections and thereby alter or reinforce users' viewpoints).[18]

> Artifacts are the means through which the human interrelates in his or her environment (i.e. they are "media") and indeed, are constitutive elements of that constantly regenerated ecology. Thus, if the "natural" *raison d'être* for making artifacts, and therefore of human techne itself, is to promote human and ecological flourishing, than the measure of flourishing itself would be the crucial criterion by which the goodness or otherwise of human creation is to be judged.[19]

In regard to digital technologies, of which AI is the contemporary paragon, the user and developer are enraptured by the technologies' enabling power as well as the appearance of spiritual – even magical – effect, and therefore imagine a definition of goodness expressed in the same information-laden terms used to describe the technology and its machines. In other words, the purpose of human nature is attributed to operating (assisting?) machines for the successful and efficient completion of "tasks" and "projects" or, for the most extreme transhumanists, the real merger of human and machine in some cyborg, science fiction fantasy. We lose a sense of the natural law that guides the faithful Christian when we stop considering the physical, spiritual, and relational composition of reality and limit our focus to gathering, manipulating, and constructing the merely virtual substance of information. Researchers are just

---

[18] Nadia Delicata, "Natural Law in a Digital Age," *Journal of Moral Theology* 4, no. 1 (2015), 7-9; Marshall McLuhan, *Understanding Media: The Extensions of Man* (McGraw-Hill, 1964).

[19] Delicata, "Natural Law in a Digital Age," 7-8.

beginning to develop an empirical and social record of these very real effects of AI; one study, for example, shows that, when workers are regularly exposed to and utilize AI applications that automate work processes, there is a significant decrease in attention to spiritual and religious faith.[20]

There is also an ideological effect associated with AI. As Neil Postman argued in *Technopoly*, "embedded in every tool is an ideological bias, a predisposition to construct the world as one thing rather than another, to value one thing over another, to amplify one sense or skill or attitude more loudly than another."[21] In the papal encyclical *Laudato si'* (which is focused on concerns about mistreatment of the environment yet has thirty-six mentions of technology), Pope Francis describes such an ideological bias as a "technocratic paradigm" that underlies the environmental and social crises.[22] Francis writes:

> This paradigm exalts the concept of a subject who, using logical and rational procedures, progressively approaches and gains control over an external object. This subject makes every effort to establish the scientific and experimental method, which in itself is already a technique of possession, mastery and transformation. It is as if the subject were to find itself in the presence of something formless, completely open to manipulation.[23]

---

[20] J. C. Jackson, K. C. Yam, P. M. Tang, C. G. Sibley, and A. Waytz, "Exposure to Automation Explains Religious Declines," *Proceedings of the National Academy of Sciences* 120, e2304748120 (2023).

[21] Neil Postman, *Technopoly: The Surrender of Culture to Technology* (New York: Vintage Books, 1993), 13.

[22] Francis, *Laudato si'* "On Care for Our Common Home" (May 24, 2015), 101-136.

[23] Francis, *Laudato si'*, 106.

What Francis appears to mean is that persons absorbed in such an ideology or persistent way of thinking are oriented mainly to their control, use, and manipulation of the world around them, including other persons. There is little attention to appreciating the goodness and nature that are already inherent in the things that are manipulated, aside from their use as objects to reach personal goals. In Francis' later document *Laudate Deum*, this ideology is explicitly connected to AI and emerging technologies, by which, he writes, "the technocratic paradigm monstrously feeds upon itself."[24] With the spread of AI throughout our culture, we might see an ever greater tendency to look upon the things and persons of this world as mere objects – tools, products, means, and parts – that are used to satisfy our self-centered desires. We might become more insensitive to the truth that our world holds meaning only from the perspective of God's wisdom, His revelation through Jesus Christ, and our human destiny in a loving relationship and eternal union with God. As Alejandro Terán-Somohano puts it, "if intelligence is reduced to computation, ethics is reduced to activism. Thus, discussions concerning the 'ethics' of AI do not ask after what is good. They are restricted, instead, to matters of control over the technology, to concerns on diversity and inclusion, privacy and transparency."[25] Philosopher David L. Schindler also worries that the American obsession with technological progress "abstracts from the logic of love proper to created being, and in so doing assumes a version of power that can only become in the end a caricature of the power of God, a power not of love but of a technical manipulation

---

[24] Francis, *Laudate Deum* "On the Climate Crisis" (October 4, 2023), 21.

[25] Alejandro Terán-Somohano, "The Banalization of Intelligence," *Word on Fire* (August 13, 2024), https://www.wordonfire.org/articles/the-banalization-of-intelligence/.

tending ultimately toward tyranny."[26] In describing "technique," which is a way of thinking and being quite similar to the technocratic paradigm, the writer Jacques Ellul puts these worries in stark terms:

> Technique worships nothing, respects nothing. It has a single role: to strip off externals, to bring everything to light, and by rational use to transform everything into means. More than science, which limits itself to explaining the "how"; technique desacralizes because it demonstrates (by evidence and not by reason, through use and not through books) that mystery does not exist. Science brings to the light of day everything man had believed sacred.[27]

We should note, of course, that the Church tradition from its earliest days has not generally been opposed to technology, including the integral development of AI as a human-centered innovation.[28] In the same address celebrating World Peace Day 2024 that is quoted above, Pope Francis stated that "we rightly rejoice and give thanks for the impressive achievements of science and technology, as a result of which countless ills that formerly plagued human life and caused great suffering have been remedied."[29] Pope Benedict XVI wrote that "technology — it is worth emphasizing — is a profoundly human reality, linked to the autonomy and

---

[26] David L. Schindler, "America's Technological Ontology and the Gift of the Given: Benedict XVI on the Cultural Significance of the Quaerere Deum," *Communio* 38.2 (2011), 237-278.

[27] Jacques Ellul, *The Technological Society* (New York: Vintage Books, 1954), 142-3.

[28] Brian Patrick Green, "The Catholic Church and Technological Progress: Past, Present, and Future," *Religions* 8, no.106, https://doi.org/10.3390/rel8060106.

[29] Francis, World Day of Peace 2024, 1.

freedom of man. In technology we express and confirm the hegemony of the spirit over matter."[30] Such assessments of technology, along with differences in its structural, motivational, and ideological effects in many different contexts, call for us to give nuanced attention to our moral evaluation of AI technology – especially concerning the consequences for our growth in virtue and our relationship to God through Christ. An important assumption in this book is that not every technology (nor every use) has the same influence on our moral behavior, and therefore we should clearly identify the relationship between each technology and moral or immoral practices as well as the mechanisms or processes and the implications associated with it.

This book explains how the nature and use of contemporary technology, through the vice of instrumental rationality, motivates sin. More specifically, this book demonstrates that artificial intelligence technology motivates acedia.[31] The vice and sin of acedia, often called sloth, received significant attention through the Middle Ages, following its fourth century designation by Evagrius Ponticus as the "most troublesome of all" evil thoughts, its traditional inclusion among the Seven Deadly Sins, and its later definition by St. Thomas Aquinas as "sorrow about spiritual good in as much as it is a Divine good" – manifesting at times as one of the most consequential mortal sins.[32] Acedia is not reducible to the simple

---

[30] Benedict XVI, *Caritas in veritate* "On Integral Development in Charity and Truth" (June 29, 2009), 69.

[31] In English, acedia commonly sounds like *ah-see-dee-ah*, but in Latin sounds like *ah-che-dee-ah*, and it derives from the Greek. For the purposes of evangelization and education, this author encourages the common pronunciation among English-speaking persons, even if the Latin pronunciation seems more "correct."

[32] Evagrius, *Praktikos*, in *The Praktikos and Chapters on Prayer*, translated by John Bamberger (Piscataway, New Jersey: Gorgias Press, 2009), 28; Thomas Aquinas, *Summa Theologica* II-II, 35, 3, trans. Fathers of the English Dominican Province (New Advent, 2017) https://www.newadvent.org/summa.

interpretation of sloth as laziness, but a long history of various interpretations of its nature and origins as well as confusion between acedia and sadness or resistance to work has left an unfortunate legacy of declining attention to its profound nature and effects. In the current age, however, concern about humanity's secularism, spiritual ennui, and struggle with a balance between work and leisure seems to have sparked renewed interest in acedia.

Instrumental rationality is a persistent disposition or vice – a habitual behavior and way of thinking (unlike "instrumental reason," which refers only to some ideal logic of thought). The person who exhibits instrumental rationality has a tendency, in many contexts, to pay less attention to good, moral, or fulfilling ends or purposes; they instead focus on choosing means that can be acquired, possessed, controlled, or used to satisfy their desires effectively and efficiently. More formally, we can define instrumental rationality as a disposition whereby a person emphasizes their successful choice of intermediate goods and means that further one or more given ends, rather than attending to the selection and guidance of appropriate ends. While instrumental rationality has not been labeled as a specific vice in Church documents, there is a clear and urgent concern about such a disposition expressed in the writings of Popes John Paul II, Benedict XVI, and Francis. Instrumental rationality, which is linked closely to a moral focus on consequences rather than principles and purposes, on purely factual and pragmatic evaluation of behavior, and an ideology of "scientism" which indicates that experimental science and practical technology are the only guides to human flourishing – essentially limiting the true ends and nature of human reason – must be challenged by a Christian perspective focused on self-giving, shared love. Benedict XVI declared that, "to resist this eclipse of reason and to preserve its capacity for seeing the essential, for seeing God and man, for seeing what is good and what is true, is the common interest that must unite all people of good will. The very future of the world is at stake."[33]

We will see that AI is a material structure and conceptual operating system that favors instrumental rationality in its operating guidelines, interpretation of data, choice of internal operations or externally oriented actions, and expectations for the human behavior of users and others. Not only does such technology generate incentives, deterrents, or constraints that result in a preference for instrumental rationality among the human persons who interact with the technology, but its proliferation and saturation in the culture will generate an environment that stifles the exercise of a holistic reason as understood in Catholic theology. The result is an environment highly conducive to humanity's engagement in acedia.

In the fifth chapter part of the book, I will provide numerous examples and information that make it abundantly clear that AI technology, through its encouragement of the vice of instrumental rationality, leads persons toward the sin of acedia. There are many evil (and good) effects and uses of AI beyond acedia, but I believe that we need to give special attention to the spread of acedia in our 21st century, hyper-technological era. The final chapter offers some concluding remarks as well as thoughts about how to remedy the sinful culture in which we find ourselves.

---

[33] Benedict XVI, Address on the Occasion of Christmas Greetings to the Roman Curia (Dec. 20, 2010), http://www.vatican.va/holy_father/benedict_xvi/speeches/2010/december/documents/hf_ben-xvi_spe_20101220_curia-auguri_en.html.

Chapter 2

The "Deadly" Sin of Acedia

The sin of acedia, one of the traditionally labeled "Seven Deadly Sins," leads at different times to apparently opposite behaviors: depressed idleness and an anxious inability to be at rest either physically or spiritually.[34] This is a profoundly distressing condition. It is also directly related to the state of the sinner's experienced relationship to God.

In the 21st century, our society and culture increasingly display widespread symptoms of acedia, or something like it. A mental health poll by the American Psychiatric Association shows that, in 2024, 43% of U.S. adults felt more anxious than they did the previous year (a number that had risen by almost a third in just two years).[35] According to the 2024 State of Mental Health in America report by Mental Health America, over 5% of the adult population and 13% of youth ages 12 to 17 had serious thoughts of suicide in 2022, and the number of suicide deaths in the U.S. was at the highest level ever recorded.[36] Aside from many additional statistics demonstrating extreme rates of anxiety, depression, loneliness, and substance abuse, there are also indications of a continued spiritual crisis; Pew Research polls show that about 28% of U.S. adults are either atheists, agnostics or "nothing in particular" regarding religion, and "when asked directly why they are not religious, two-thirds of 'nones' say they question

---

[34] See footnote 17 regarding pronunciation.

[35] American Psychiatric Association, "American Adults Express Increasing Anxiousness in Annual Poll; Stress and Sleep are Key Factors Impacting Mental Health" (May 1, 2024), https://www.psychiatry.org/news-room/news-releases/annual-poll-adults-express-increasing-anxiousness.

[36] Mental Health America, *State of Mental Health in America* (2024), https://mhanational.org/issues/state-mental-health-america.

a lot of religious teachings or don't believe in God."[37] According to Gallup, only about 3 in 10 American adults, including Catholics, attend church almost weekly.[38]

We cannot, of course, demonstrate with statistics that Americans are engaged in the sin of acedia. This requires a deep dive into the intentions of persons that will not be attempted here. With a clear understanding of acedia, however, most Christians will be able to recognize its prevalence among friends, family, and peers and throughout our society. That is the goal of this chapter. We will still need to link acedia to the motivational influence of AI; that will be the task of the following chapters.

Acedia's Introduction

Originally, the ancient Greek word ακεδια had a vague meaning of "lack of care" that could be applied in a variety of contexts.[39] As a term specifically related to moral and religious concerns, however, acedia first became known in the fourth century writings of Evagrius Ponticus. He was a highly influential Christian theologian who most notably lived a very simple life of physical hardship and intense prayer among ascetic (self-denying) communities in Egypt.

Evagrius did not precisely define acedia but instead described the many ways it is experienced:

---

[37] Pew Research, "Religious 'Nones' in America: Who They Are and What They Believe" (January 24, 2024), https://www.pewresearch.org/religion/2024/01/24/religious-nones-in-america-who-they-are-and-what-they-believe/.

[38] Jeffrey M. Jones, "Church Attendance Has Declined in Most U.S. Religious Groups," Gallup (March 25, 2024), https://news.gallup.com/poll/642548/church-attendance-declined-religious-groups.aspx.

[39] Siegfried Wenzel, *The Sin of Sloth: Acedia in Medieval Thought and Literature* (Chapel Hill: University of North Carolina Press, 1967), 6.

Acedia is an ethereal friendship, one who leads our steps astray, hatred of industriousness, a battle against stillness, stormy weather for psalmody, laziness in prayer, a slackening of [self-discipline], untimely drowsiness, re-volving sleep, the oppressiveness of solitude, hatred of one's cell, an adversary of ascetic works, an opponent of perseverance, a muzzling of meditation, ignorance of the scriptures, a partaker in sorrow, a clock for hunger.[40]

The agitation of a person afflicted with acedia could often be experi-enced as restlessness that "instills in him a dislike for the place and for his state of life itself."[41]

The eye of the person afflicted with acedia stares at the doors continuously, and his intellect imagines people coming to visit. The door creaks and he jumps up; he hears a sound, and he leans out the window and does not leave it until he gets stiff from sitting there.

When he reads, the one afflicted with acedia yawns a lot and readily drifts off into sleep; he rubs his eyes and stretches his arms; turning his eyes away from the book, he stares at the wall and again goes back to reading for awhile; leafing through the pages, he looks curiously for the end of texts, he counts the folios and calculates the number of gatherings.

---

[40] Evagrius Ponticus, "On the Vices Opposed to the Virtues" 4, in *Evagrius of Pontus: The Greek Ascetic Corpus*, trans. Robert E. Sinkewicz (Oxford: Oxford University Press, 2006), 64.

[41] Evagrius Ponticus, *Praktikos*, in *The Praktikos and Chapters on Prayer*, trans. John Bamberger (Piscataway, New Jersey: Gorgias Press, 2009), 12; Jeffrey A. Vogel, "The Speed of Sloth: Reconsidering the Sin of Acedia," *Pro Ecclesia* 18, no. 1, 60-61.

Later, he closes the book and puts it under his head and falls asleep, but not a very deep sleep, for hunger then rouses his soul and has him show concern for its needs.[42]

Evagrius' writings express his belief that the core of a holy life is an effort to attain *apatheia* (ἀπάθεια), a peaceful state in which the passions – we might loosely call them emotions – are fully integrated with the rational part of the soul. Evagrius advocated for a psychological model of human persons, drawn from the philosopher Plato, in which the soul is comprised of distinct aspects: the concupiscible (bodily desires), irascible (energy and emotions), and rational. When passions are said to be integrated with the rational part of the soul, it means that, even though a person may feel passions deeply, they are not in conflict with the thoughts and actions that are truly reasonable. Unfortunately, few persons succeed in integrating their passions so well, and various passions, arising either in the body or soul, tempt the soul away from virtue.

The temptations arise when a person engages in certain thoughts. "It is easier to sin by thought than by deed... For the mind is easily moved indeed, and hard to control in the presence of sinful fantasies."[43] There are, in particular, eight thoughts that lead a person toward evil in a fixed conceptual order: gluttony, sexual impurity, avarice, sadness, anger, acedia, vainglory, and pride.[44] Acedia is most closely related to sadness and anger, although it is distinct and tends to follow from the others.[45] The foundation for all of these dangerous thoughts is love of self which is the

---

[42] Evagrius Ponticus, *On the Eight Thoughts*, in *Evagrius of Ponticus: The Greek Ascetic Corpus*, trans. Robert E. Sinkewicz (Oxford: Oxford University Press, 2003), 84.

[43] Evagrius, *Praktikos*, 48.

[44] Evagrius, *Praktikos*, 6-14.

[45] Mark Sultana, "Combatting Acedia: The Neptic Antidote," *Heythrop Journal* 63, no. 4 (2019), 835.

contrary of virtuous love.[46] The primary virtues in a holy life are then love and self-control, each of which helps to keep anger and desire in balance with more rational inclinations.[47]

The role of these thoughts in a person's sinfulness is crucial to understanding the essential freedom of choice involved. "It is not in our power to determine whether we are disturbed by these thoughts, but it is up to us to decide if they are to linger within us or not and whether or not they are to stir up our passions."[48] The thoughts themselves are not sins or vices, but they can be experienced as "evil" because they tend toward passionate excesses and the vices that the inordinate passions engender. For example, a thought like the love of one's family is quite compatible with the goodness of human nature, yet it may be perverted when a person treats loyalty to family as if it is above fidelity to God's law. Anger, which is a natural faculty oriented to fighting for desired goods, can also become distorted and instead be an obstacle to the good of *apatheia* and knowledge.[49]

There is a further element in the temptation of human persons, and that is the constant presence of demons that encourage, argue, deceive, or otherwise motivate people to entertain the kinds of thoughts that will lead to inordinate passions and the subsequent vices. Evagrius referred to acedia in its personalized form as "the noonday demon." Demons and the devil are, for Evagrius, very real, personal manifestations of evil in the world. To be clear, evil is not an eternal force opposed to good but a privation of the good – an absence of good – that reveals itself in behavior through individuals' choices to pervert their natural goodness. Even "the

---

[46] Sultana, "Combatting Acedia," 833.

[47] Gabriel Bunge, *Despondency: The Spiritual Teaching of Evagrius of Pontus*, trans. Anthony P. Gythiel (Yonkers, NY: St. Vladimir's Seminary Press, 2011), 30.

[48] Evagrius, *Praktikos*, 6.

[49] Kathleen S. Gibbons, "Vice and Self Examination in the Christian Desert," PhD dissertation (Toronto: University of Toronto, 2011), 187-189.

devil is not evil by nature," but distorts his own goodness through choice and action.[50]

> There was a time when evil did not exist, and the time will come when it will exist no more. However, there was never a time when virtue did not exist and there will be no time when it will no longer be. For the seeds of virtue are imperishable. The rich man [in the New Testament] convinces me of this; he was in hell because of his malice yet took pity on his brothers. But having compassion is an excellent seed of virtue.[51]

In Evagrius' moral psychology, the demons must be fought by human persons because of the demons' persistent and intentional temptations and due to the weakness (instability) of Christians who are still on the path toward a life anchored in virtue and holiness. This battle becomes especially difficult as demons find opportunities to direct their efforts toward the specific weaknesses and characters of the individuals who are besieged; here, not only human nature but the personal nature of the individual is perverted in the attempt to manifest evil. Ultimately, the outcome of the battle is within the control of the person, for the demons are not any more determinant of the person's thoughts than the sensations, memories, or moods that each individual encounters daily.[52]

Evagrius' ultimate interest is expressed as an intellectual goal: to perceive and understand one's divine purpose. "Both the virtues and vices

---

[50] Evagrius Ponticus, *Kephalaia Gnostika: A New Translation from the Unreformed Text from the Syriac.* Translated by Ilaria L. E. Ramelli (Atlanta, SBL Press, 2015), 4, 59; cited by Bunge, *Despondency,* 19.

[51] Evagrius Ponticus, *Kephalaia Gnostika,* 1, 40; cited by Bunge, *Despondency,* 31.

[52] Bunge, *Despondency,* 21.

make the man blind. The one so that it may not see the vices; the other, in turn, so that it might not see the virtues."[53] The "evil" thoughts that tend toward vice are essentially mental failings caused by faithlessness, fear, or ignorance; they arise out of the concupiscible and irascible aspects of the soul, potentially unseating the dignity of the rational aspect. "The spirit that is engaged in the war against the passions does not see clearly the basic meaning of the war for it is something like a man fighting in the darkness of night. Once it has attained purity of heart though, it distinctly makes out the designs of the enemy."[54] *Apatheia* is therefore not an absence of desire, but a calming of inordinate passions and healthy integration of the three aspects of the soul in order to attain clear knowledge. "The one whose mind is always with the Lord and whose incensive part is full of humility through remembrance of God, and whose entire desire is inclined to the Lord, has no need to be afraid of our enemies who encircle our body on the outside."[55]

It is not merely the state of *apatheia* that is the proper end of a person in the moral life, but its further enabling of active contemplation of God and His creation.[56] For Evagrius, the ultimate goal in a holy life is attainment of mystical knowledge of the Divine (going "immaterially to the Immaterial") in a final stage of achievement called *gnostikē*.[57] This is a profound kind of contemplation of all creation and of God that

---

[53] Evagrius, *Praktikos,* 62; Brandon Dahm, "Correcting Acedia through Wonder and Gratitude," *Religions* 12, no. 7 (2021), 459.

[54] Evagrius, *Praktikos*, 83.

[55] Evagrius, *Kephalaia Gnostika*, 4, 73; cited by Bunge, *Despondency*, 32.

[56] Robert W. Daly, "Before Depression: The Medieval Vice of Acedia," *Psychiatry* 70, no. 1 (Spring 1970), 33.

[57] Christopher D. Jones, "The Problem of Acedia in Eastern Orthodox Morality," *Studies in Christian Ethics* 33, no,3 (2019), 337-8.

transforms the mind so it may reflect on non-sensible (immaterial) reality.[58] This is not simply a matter of human effort, as the heretical Gnostics believed, for it is cooperation with divine grace, a gift offered by God, that generates *synergeia* and a harmony of wills with the Holy Spirit.[59]

The "noonday demon" incites acedia in the vulnerable person with a comprehensive assault on all parts of the soul that potentially undermines progress toward *gnostikē*. This demon is particularly bold in its daylight activity, and it is described as an army commander that "deploys every device in order to have the monk leave his cell."[60] Among the eight "evil" thoughts, acedia is "the most troublesome of all" and can be spiritually catastrophic.[61] The experience of acedia is "mixed, coming to us both as animals and as human beings,"[62] but is particularly dangerous because it leads to overcoming the rational power which surrenders in its thoughts to the irrational, animal parts of the soul. Jeffrey Vogel refers to acedia as the "breakdown of the spiritual immune system" by which the person is made vulnerable to the more spiritual and more serious thoughts of vainglory and pride.[63] J. L. Aijian summarizes the challenges well: "ακεδια is an ethereal friendship, one who leads our steps astray, hatred of industriousness, a battle against stillness, stormy weather for psalmody, laziness in prayer, a slackening of ascesis, untimely drowsiness, revolving sleep,

---

[58] Evagrius, *Kephalaia Gnostika*. "Just as the senses are changed by the perceptions of different qualities, likewise the intellect too is changed, when it meditates on contemplations that are different every moment," 2, 83. "Contemplation is the spiritual knowledge of those realities that were and will be, which lifts the intellect up toward its original order," 3, 42.

[59] Bunge, *Despondency*, 42.

[60] Evagrius, *Praktikos*, 12.

[61] Evagrius, *Praktikos*, 28.

[62] Evagrius Ponticus, *Reflections*, 40, in *Evagrius of Pontus,* trans. Sinkewicz; J. L. Aijian, "Fleeing the Stadium: Recovering the Conceptual Unity of Evagrius' Acedia," *Heythrop Journal* 62, no.1, 11.

[63] Vogel, "The Speed of Sloth," 60.

the oppressiveness of solitude, hatred of one's cell, an adversary of ascetic works, an opponent of perseverance, a muzzling of meditation, ignorance of the scriptures, a partaker in sorrow, a clock for hunger."[64]

Acedia is also unlike the other evil thoughts in its tendency to endure rather than mediate the transition to other thoughts. This sequential finality of acedia underscores the need for perseverance and vigilance in overcoming it: "Set a measure for yourself in every work and do not let up until you have completed it. Pray with understanding and intensity, and the spirit of acedia will flee from you."[65] Such overcoming is therefore the end of a sequence or process of purification that is all the more worthy of celebration when achieved. As the main obstacle to contemplative *gnostikē*, acedia's challenge poses an opportunity for victory over the fundamental essence of vice itself: self-love.

In the 5th century, St. John Cassian communicated the wisdom about acedia to the West. He was particularly concerned with providing virtuous advice to new monastic communities, which differed from the communities in which Evagrius lived by their greater emphasis on communal activities, hierarchy of positions, and organized work schedules. St. Cassian did, however, spend some time among the Desert Fathers and appears to have absorbed Evagrius' teachings about acedia. His descriptions of acedia resemble Evagrius' litany of causes and effects, with perhaps more emphasis on the emotional anxiety of the afflicted persons and a significant elevation of the importance of manual labor as a form of perseverance, communal solidarity (as for the monk), and resistance to temptation.[66] He claims that "all the inconveniences of this disease are

---

[64] Evagrius Ponticus, "On the Vices Opposed to the Virtues" 4.

[65] Evagrius Ponticus, "On the Eight Thoughts," 18, in *Evagrius of Pontus,* trans. Sinkewicz.

[66] Dennis Ockholm, "Staying Put to Get Somewhere," in *Acedia,* ed. Robert B. Kruschwitz (Waco, Texas: Center for Christian Ethics at Baylor University, 2013), 23.

admirably expressed by [King] David in a single verse, where he says, 'My soul slept from weariness,' that is, from acedia."[67]

Work is not the only prescription, however, for many acts of endurance, courage, prayer, and charity are effective in countering the harmful effects of acedia. The communal disruption caused by both inordinate busyness and sloth or laziness is particularly important to moral evaluation of a monk's behavior, and anxiety "makes a person horrified at where he is, disgusted with his cell, and also disdainful and contemptuous of the brothers."[68] According to Amy Freeman, St. Cassian suggests that it is the turn away from diligent labor that typically leads a person to acedia, with the anxious expressions of restlessness or busyness following (based on an interpretation of 2 Thessalonians 3:6-15 that explicitly ties an acedia-like condition to idleness, disquietude, and meddling).[69] Some anxieties, however, can encourage resignation and idleness:

> We lament that in all this while, living in the same spot, we have made no progress, we sigh and complain that bereft of sympathetic fellowship we have no spiritual fruit; and bewail ourselves as empty of all spiritual profit, abiding vacant and useless in this place, and we that could guide others and be of value to multitudes have edified no man, enriched no man with our precept and example.[70]

---

[67] Cassian, *Institutes*, trans. C.S. Gibson, ed. Philip Schaff and Henry Wace, rev. and ed. for New Advent by Kevin Knight, in *Nicene and Post-Nicene Fathers*, 2nd Series, Vol. 11 (Buffalo, New York: Christian Literature Publishing Co., 1894), 10, 4. http://www.newadvent.org/fathers/3507.htm.

[68] Cassian, *Institutes*, 10, 2.

[69] Amy Freeman, "Remedies to Acedia in the Rhythm of Daily Life," in Kruschwitz, *Acedia*, 36; Cassian, *Institutes*, 10, 15.

[70] Cassian, *Institutes*, 10, 2.

Acedia may arise from any combination of vices, but especially from sadness, which it follows in the progression of the eight vices. Sadness can have its origin in "unreasonable mental anguish or from despair" and also in the aftermath of the person's experience of anger, hurt, or unfulfilled desire.[71] Neither acedia nor any of the vices are limited to the experiences of one particular group of persons, for "they do not assail everyone in the same way. ... We suffer in different ways and manners."[72]

St. Thomas Aquinas' Theological Foundation for Acedia

In St. Thomas Aquinas' thirteenth century writings, acedia is defined as "sorrow about spiritual good in as much as it is a Divine good."[73] Sorrow in general arises when a person encounters a good – something that is truly desirable – or an evil.[74] Its cause is a current lack of possession or attainment of the desired good, precisely the opposite of satisfied rest of the person's appetite in an already obtained good, which is experienced as joy.[75] With acedia in particular, the afflicted person feels a current loss of the Divine good, but that person struggles against the resignation, anxiety, restlessness, and doubt (a current loss of trust, security, peace, faith, etc.) that cause the physically and mentally experienced sorrow.

Acedia is an evil only partly because it is a kind of sorrow. Because it is a form of pain, we can, in a sense, call sorrow evil.[76] It is the object (goal or focus) of the sorrow, however, that makes it morally good or evil.[77]

---

[71] Cassian, *Conferences*, 5, 11.

[72] Cassian, "On the Eight Deadly Sins," 194.

[73] Thomas Aquinas, *Summa Theologica* II-II, 35, 3, trans. Fathers of the English Dominican Province (New Advent, 2017) https://www.newadvent.org/summa.

[74] Aquinas, *Summa Theologica* I-II, 35, 2-3.

[75] Aquinas, *Summa Theologica* I-II, 35, 3.

[76] Aquinas, *Summa Theologica* I-II, 39, 1.

[77] Aquinas, *Summa Theologica* I-II, 35, 6.

When a person applies reason to their command of or willed consent to their passions, then that person is truly responsible for their behavior, and moral good or evil is assigned. Evil is therefore a moral attribution properly applied only when use of the power of reason is also involved.[78] When a rational person experiences sorrow over a good, it is contrary to the moral and objective good of the person and therefore evil.[79] It is reason's consent with such a sorrow that would make it a moral evil, for "man is reckoned to be good or evil chiefly according to the pleasure of the human will."[80]

The key to identifying the evil in acedia is found in the afflicted person's sorrowful experience of its object. St. Thomas' clearest and fullest expression of this object is found in his description of acedia as "sorrow about spiritual good in as much as it is a Divine good."[81] Here, the most obvious definition of the "Divine good" would be God Himself, but St. Thomas explains that "acedia is not sadness about the presence of God himself, but sadness about some good pertaining to him which is divine by participation."[82] This is because it is not possible to be sorrowful about God Himself when communion with God is a person's true end and beatitude.[83] What may be possible, however, is that the "Divine good" to which acedia is oriented can be a person's enjoyment of God's love, for "man's last end may be said to be God who is the Supreme Good simply; or it can be man's enjoyment of God, which implies a certain pleasure in

[78] Aquinas, *Summa Theologica* I-II, 24, 1.

[79] Aquinas, *Summa Theologica* I-II, 36, 1.

[80] Aquinas, *Summa Theologica* I-II, 34, 4; 39, 3, ad 2.

[81] Aquinas, *Summa Theologica* II-II, 35, 3.

[82] Thomas Aquinas, *Questiones Disputatae de Malo* (On Evil) 11, 3, ad 3, trans. Richard Regan (Oxford: Oxford University Press, 2003).

[83] Thomas Aquinas, *Summa Contra Gentiles* 1, 89, 9, trans. Charles J. O'Neil (Garden City, New York: Image Books, 1957).

the last end."[84] Enjoyment here seems to refer not only to delight in God but also the personal experience of a relationship with God in time and place. When St. Thomas states that acedia is sorrow about the spiritual and Divine good, which is found in one's experience of God's love, he is not contradicting himself by indicating that a person can have both sorrow and joy over precisely the same object. The potential joy of a person afflicted with acedia is in God's love taken absolutely, but their sorrow is associated with their personal experience of that love. It seems that such a person's joyful experience of that love is diminished, or even lost.

St. Thomas declares, in fact, that acedia is the sorrow opposed to the joy of charity. He is referring here to the joy associated with the virtue of charity, which as a theological virtue is dependent on the grace of God for its presence and development in the human person.

> There can be spiritual joy about God in two ways. First, when we rejoice in the Divine good considered in itself; secondly, when we rejoice in the Divine good as participated by us. The former joy is the better, and proceeds from charity chiefly: while the latter joy proceeds from hope also, whereby we look forward to enjoy the Divine good, although this enjoyment itself, whether perfect or imperfect, is obtained according to the measure of one's charity.[85]

Acedia is the sorrow which undermines, decreases, and is experienced as the contrary passion to the joy that is associated with charity. It seems that a person can fail in a limited way to perceive or understand God's offer of love and communion, thereby experiencing self-induced

---

[84] Aquinas, *Summa Theologica* I-II, 34, 3

[85] Aquinas, *Summa Theologica* II-II, 28, 1, ad 3.

sorrow over a perceived evil. The intense desire for God's love (whether fully conscious or psychologically sublimated) coupled with a frustrated experience of that love can be a source of profound sorrow, grief, and despair: "Since desire, the first effect of love, gives rise to the greatest delight or pleasure, desire is also the universal cause of sorrow – of withdrawal or flight from all that is opposed to such delight."[86]

Charity is the friendship of persons for God, moved by the Holy Spirit rather than by the human will on its own.[87]

> [T]he act of charity surpasses the nature of the power of the will, so that, therefore, unless some form be superadded to the natural power, inclining it to the act of love, this same act would be less perfect than the natural acts and the acts of the other powers; nor would it be easy and pleasurable to perform. And this is evidently untrue, since no virtue has such a strong inclination to its act as charity has, nor does any virtue perform its act with so great pleasure. Therefore it is most necessary that, for us to perform the act of charity, there should be in us some habitual form superadded to the natural power, inclining that power to the act of charity, and causing it to act with ease and pleasure.[88]

Therefore, even a person's charitable actions are dependent on the initial infusion of charity into his soul. It is not possible for a human person to increase or decrease charity in themselves by depending on their own

---

[86] Aquinas, *Summa Theologica* ST, I-II, 36, 2.
[87] Aquinas, *Summa Theologica* II-II, 23, 2.
[88] Aquinas, *Summa Theologica* II-II, 23, 2.

powers, "but only on the will of the Holy Spirit, who divides his gifts according to his will."[89]

The will of the person, however, is also involved in the theological virtue of charity, since the "Divine good" is apprehended and desired in the person's intellect.[90] The human will is therefore both moved and moves in charity. The friendship with God requires not only a mutual benevolence and active loving, but communication, for "this well-wishing is founded on some kind of communication."[91] This "conversation" is communication of God's happiness, imperfectly experienced by a person in their spiritual nature. Although each person must love all that pertains to God with their whole being, it is not possible to love God "wholly" in the same way that God loves.

> Just as God is supremely knowable in Himself yet not to us, on account of a defect in our knowledge which depends on sensible things, so too, God is supremely lovable in Himself, in as much as He is the object of happiness. But He is not supremely lovable to us in this way, on account of the inclination of our appetite towards visible goods. Hence it is evident that for us to love God above all things in this way, it is necessary that charity be infused into our hearts.[92]

Therefore, although a person in friendship with God is in a loving relationship, and so loves God immediately yet finitely, the person's active loving may be more or less perfect in its operation and in the extent to

---

[89] Aquinas, *Summa Theologica* II-II, 24, 3.

[90] Aquinas, *Summa Theologica* II-II, 24, 3.

[91] Aquinas, *Summa Theologica* II-II, 23, 1.

[92] Aquinas, *Summa Theologica* II-II, 24, 2, ad 2.

which they love God in their whole being. The person infused with the virtue of charity is helped by the Holy Spirit's gift of wisdom, whereby the person may know and love Divine truth beyond the capacity of reason. "This sympathy or connaturality for Divine things is the result of charity, which unites us to God."[93]

St. Thomas further indicates that "the joy of charity is joy about the Divine wisdom."[94] This joy is not merely "extra" or superfluous to the Divine good, as some passions may follow or derive from a moral thought or action; rather, happiness is itself a human person's one end in life.[95] A person's happiness and enjoyment of it is found in the universal good, which is God.[96] In this life, our incomplete participation in happiness is undermined by ignorance, inordinate and excessive passions, and bodily temptations and pains.[97] On the other hand, a person's joy in charity can be enhanced by knowing and willing his end in God; human nature "did give him free-will, with which he can turn to God that He may make him happy. For what we do by means of our friends, is done, in a sense, by ourselves (Ethic, iii.3)."[98] St. Thomas makes this point clearly, echoing St. Augustine: "joy is the volition of consent to the things we wish."[99] We must cooperate with the gifts of the Holy Spirit.

Because acedia is the sorrow that opposes the joy of charity, it is not merely a passive sorrow. The sinner willfully turns away the Divine good as it is participated in by human beings. This aversion from God is not hatred, for it is not directly an aversion to God (which is not precisely

---

[93] Aquinas, *Summa Theologica* II-II, 45, 2.

[94] Aquinas, *Summa Theologica* II-II, 28, 2.

[95] Aquinas, *Summa Theologica* II-II, 28, 2.

[96] Aquinas, *Summa Theologica* I-II, 2, 8.

[97] Aquinas, *Summa Theologica* II-II, 28, 3.

[98] Aquinas, *Summa Theologica* I-II, 5, 5, ad 1.

[99] Aquinas, *Summa Theologica* I-II, 31, 4; Augustine, *The City of God* 14, 6, trans. Marcus Dods (New York: The Modern Library, 2000).

possible) or the effects of God,[100] but an aversion to the experience of joy over our participation in those effects because they are somehow experienced as distasteful or burdensome. Acedia, in its mortal form, is therefore a kind of response to a spiritual weariness that inspires "dislike, horror and detestation of the Divine good, on account of the flesh utterly prevailing over the spirit."[101] It is a refusal of the Divine good – not of God in Himself, or of His love, but of enjoyment of God's love, goodness, truth, and more.

St. Thomas indicates that "acedia is a mortal sin by reason of its genus."[102] A mortal sin is one that substantially undermines the relationship of a person with God and can only be repaired through God's free gift of grace, while a venial sin also harms the sinner's relationship with God but is not irreparable. In its spiritual form acedia is a mortal sin, because it is directly opposed to the joy of charity.[103] There is a potential continuity between venial and mortal acedia, as when the sinner, "through accustoming his will not to be subject to the due order in lesser matters, is disposed not to subject his will even to the order of the last end, by choosing something that is a mortal sin in its genus."[104] Here, we see the influence of a persistent worry that runs all the way back to Evagrius, who warned that "it is easier to sin by thought than by deed… For the mind is easily moved indeed, and hard to control in the presence of sinful fantasies."[105]

The experience of human persons occurs in bodily sensations as well as spiritual movements, and this is reflected in St. Thomas' varying references to acedia in its carnal and spiritual manifestations. It can be initially confusing to read that "acedia is a kind of sorrow, whereby one becomes

---

[100] Aquinas, *Summa Theologica* II-II, 34, 1.
[101] Aquinas, *Summa Theologica* II-II, 35, 3.
[102] Aquinas, *Summa Theologica* II-II, 35, 3.
[103] Aquinas, *Summa Theologica* I-II, 88, 1.
[104] Aquinas, *Summa Theologica* I-II, 88, 3.
[105] Evagrius, *Praktikos* 48.

sluggish in spiritual acts because they weary the body"[106] and it is "sadness about one's spiritual good, on account of the attendant bodily labor,"[107] yet it is clearly defined as "sorrow about spiritual good in as much as it is a Divine good."[108] What does bodily labor or fatigue have to do with the Divine good?

It is important to note that St. Thomas describes acedia as both a vice and a sin, and as a sin it can be either venial or mortal. In an important passage, he teaches that:

> [T]he movement of sloth is sometimes in the sensuality alone, by reason of the opposition of the flesh to the spirit, and then it is a venial sin; whereas sometimes it reaches to the reason, which consents in the dislike, horror and detestation of the Divine good, on account of the flesh utterly prevailing over the spirit. On this case it is evident that sloth is a mortal sin.[109]

In both its venial and mortal forms, then, acedia involves an encounter with desires of "the flesh." As a venial sin, such fleshly concerns are found "in the sensuality alone," and this suggests bodily or basic psychological temptations that generate avoidance of pain or satisfaction of such limited urges. Weakness or ignorance are causes of sins associated with the sensitive (sense-driven) appetite.[110] This "opposition of the flesh to the spirit" is then an irrational or immoderate surrender of the will to one's sensitive appetite, and it is sinful for this usurpation of reason, yet

---

[106] Aquinas, *Summa Theologica* I, 63, 2, ad 2.

[107] Aquinas, *Summa Theologica* I-II, 84, 4.

[108] Aquinas, *Summa Theologica* II-II, 35, 3.

[109] Aquinas, *Summa Theologica* II-II, 35, 3.

[110] Aquinas, *Summa Theologica* I-II, 88, 2.

classified as venial because it is not instigated by the will against the Divine good of friendship with God.

A person's struggle with disordered temptations of sensual urges and avoidances, however, does not remove this Divine good as the spiritual object of acedia; the passage quoted above specifies the differences between acedia as a venial and mortal sin, but it does not indicate that the Divine good is only the object of acedia as a mortal sin and kind of sorrow. Acedia as a venial sin concerns the carnal desires as impediments in attaining the good of friendship with God, and the sinner therefore sorrows about this spiritual good which is, in their experience, opposed to the pleasures of succumbing to particular, carnal impulses of the sensitive appetite. "For since the pleasures of the flesh concern goods which are beneath us, whereas the pleasures of the spirit concern goods which are above us, it comes to pass that when the soul is occupied with the lower things of the flesh, it is withdrawn from the higher things of the spirit."[111] Without willfully casting off the "yoke" of this Divine friendship, the venial sinner engages in acedia by allowing or enabling a false conflict between their sensual drives and their true end. "And so when desire of the flesh is dominant in human beings, they have distaste for spiritual good as contrary to their good. Just so, human beings with infected taste buds have distaste for healthy food and grieve over it whenever they need to consume such food."[112] This is not, however, an intractable conflict of bodily and soulful natures within the sinner, for "the spirit does not lust against the nature of the flesh, but against its desires, namely, those that concern superfluities."[113]

---

[111] Thomas Aquinas, *Commentary on St. Paul's Letter to the Galatians* 5, 4, 12, trans. F. R. Larcher (Albany: Magi Books, 1966), https://isidore.co/aquinas/english/SSGalatians.htm.

[112] Aquinas, *De Malo*, 11, 2.

[113] Aquinas, *Commentary on Galatians* 5, 4.

The key difference in acedia as a mortal sin is not the new role of "the flesh," but the substitution of "reason" for "the sensuality alone" in characterizing sinful action. In the passage quoted above from St. Thomas' *Summa Theologica* II-II, q.35, a.3, the structure of the sentence contrasts sensuality and reason; the "opposition" or "prevailing" of the flesh, on the other hand, is indicated as a secondary motivation but not the essence of each respective form of sin. Therefore, when we interpret the meaning of the "opposition" of flesh and spirit in venial acedia, we must understand it as an opposition in the sensuality alone: a carnal temptation or obstacle. In regard to acedia as a mortal sin, on the other hand, the "prevailing" of the flesh over the spirit must be something that is also spiritual, engaging the intellectual appetite. Unlike the venial sin of acedia, the mortal sin of acedia cannot merely be due to "the concupiscence that occurs in sensuality alone."[114] St. Thomas would not suggest that the flesh, understood reductively as the purely carnal appetite, would ever "prevail" over the spirit in a continent and morally responsible person. A human person is essentially a spiritual animal whose soul is characterized by spiritual powers, even if a person may – with free will – relinquish temporary control to the lower powers and appetite.

It is helpful that St. Thomas elsewhere explains that a person "is said to live according to the flesh, when he lives according to himself, as Augustine says (De Civ. Dei xiv, 2,3). The reason of this is because every failing in the human reason is due in some way to the carnal sense."[115] In his *Commentary on St. Paul's Letter to the Galatians*, St. Thomas further indicates that:

> [A] sin can be said to be "of the flesh" in two ways:
> first, sins are "of the flesh" that are fulfilled in the pleasure

---

[114] Aquinas, *De Malo* 11, 3, ad 8.

[115] Aquinas, *Summa Theologica* I-II, 72, 2, ad 1.

of the flesh, as are gluttony and lust. Or, sins are "of the flesh" with respect to their root: in this sense all sins are called sins of the flesh, inasmuch as the soul is so weighed form by the weakness of the flesh (as is written in Wisdom 9:15) that the enfeebled intellect can be easily misled and hindered from operating perfectly.[116]

Weakness in this context is an experienced decrease in capacity of the soul, analogous to physical weariness.

The prevailing of flesh over spirit in mortal acedia is also not merely a generic kind of spiritual malady but is instead sorrow over a Divine good in particular. "Sloth is not an aversion of the mind from any spiritual good, but from the Divine good, to which the mind is obliged to adhere."[117] The prevailing of self-regarding "flesh" over spirit takes on a new moral gravity when it concerns this Divine good. It seems that the means of prevailing has to do with negligence in knowledge or awareness of the Divine good that is subordinated in action, for "all the sins which are due to ignorance, can be reduced to sloth, to which pertains the negligence of a man who declines to acquire spiritual goods on account of the attendant labor; for the ignorance that can cause sin, is due to negligence ...."[118]

This is an essentially spiritual disorder, both a "lack of knowledge of those things that one has a natural aptitude to know"[119] and a willed refusal to acknowledge the final good of human persons in friendship with God. One not only ought to know such a principle – one is obligated to do so as an article of faith.

---

[116] Aquinas, *Commentary on Galatians* 5, 5.
[117] Aquinas, *Summa Theologica* II-II, 35, 3, ad 2.
[118] Aquinas, *Summa Theologica* I-II, 76, 2.
[119] Aquinas, *Summa Theologica* I-II, 76, 2.

As a capital vice, acedia "easily gives rise to others, as being their final cause."[120] St. Gregory had listed seven "daughters" of acedia, and St. Thomas repeats these as "malice, spite, faint-heartedness, despair, sluggishness in regard to the commandments, [and] wandering of the mind after unlawful things."[121] He further indicates that the additional daughter vices identified by St. Isidore of Seville – bitterness, idleness, drowsiness, uneasiness of the mind, curiosity, loquacity, restlessness of the body, and instability – are implied by St. Gregory's labels.[122] Rebecca Konyndyk DeYoung points out that St. Thomas finds an equivalent final cause in pursuit of a good and avoidance of the contrary evil; "[acedia] itself is better characterized as the impulse to escape something burdensome than as something good one desires as a convenient escape or diversion."[123] With acedia, of course, the object of sorrow is burdensome only in a way that is either tangential to one's final end or falsely understood as an evil; "just as we do many things on account of pleasure ... so again we do many things on account of sorrow, either that we may avoid it, or through being exasperated into doing something under pressure thereof."[124]

[120] Aquinas, *Summa Theologica* II-II, 35, 4.

[121] Aquinas, *Summa Theologica* II-II, 35, 4, obj 2.

[122] Aquinas, *Summa Theologica* II-II, 35, 4, ad 3.

[123] Rebecca Konyndyk DeYoung, "Aquinas on the Vice of Sloth: Three Interpretive Issues," *The Thomist* 75, no. 1 (2011), 48.

[124] Aquinas, *Summa Theologica* II-II, 35, 4.

Chapter 3

The Vice of Instrumental Rationality

This part of the book concerns the nature of instrumental rationality, its importance to moral teaching in the writings of recent popes and in the Bible, and its relation to the sin of acedia and to AI. Chapter 3 specifies the differences between instrumental reason and rationality and provides a more formal definition of instrumental rationality. The chapter explains how instrumental rationality is linked to acedia. Chapter 4 examines the many references to instrumental rationality or related themes in the teaching of Popes John Paul II, Benedict XVI, and Francis. These are the three popes who have addressed instrumental rationality most extensively, and a review of their teachings offers a fairly thorough explanation of the associated moral problems. Several portions or themes of the Bible, found in both the Old and New Testaments, are interpreted as sources of principles that may help us to evaluate instrumental rationality. Finally, in Chapter 4, I go into some detail regarding ways in which instrumental rationality is structurally represented, and then encouraged and even enforced, through AI technology.

Instrumental Rationality as a Problem

Instrumental rationality has to do with the calculation and choice of means to attain deliberate ends or goals, but it is much more than that, especially when we evaluate it from a moral perspective. In the theological, philosophical, and popular literature, there is a wide variety of interpretations of instrumental rationality as well as a tendency to use the term without any explicit definition. Moreover, there is often

little intentional thought about the assumptions behind it and their consequences. In order to more clearly define instrumental rationality here,
we will need to draw on the classic foundation in Max Weber's works as
well as form distinctions that are rarely made outside of the social sciences.

For the purpose of clarity, we should establish boundaries between
the concepts of instrumental reason, reasoning, and rationality. In comparison to reason in general, *instrumental reason* is a human power of
accessing truth when intentionally pursuing pragmatic interests that
link means to ends. The "truth" involved here is typically understood as
something very different than what is associated with *practical reason*
in Catholic theology and philosophy, for the truth of practical reason
has to do with action in general, and it guides that action to our real,
naturally indicated goals and purposes – ultimately to the end of a loving relationship with God, our final or supreme end. The truth of practical reason can be expressed simply as the intellectual light that leads a
person to do good. The "truth" of instrumental reason, on the other
hand, is more limited; it is focused on the attainment of a specified goal
that is desired and intended by a person. In many cases it is a factual
truth, describing the real-world tools and obstacles involved in actually
reaching a goal. For most people who use the term, instrumental reason
is oriented toward a truth that is decidedly *not* moral or tied to concerns
like beauty, integrity, passion, etc. That does not mean that a person is
necessarily committing immoral, evil acts because they use so-called
instrumental reason, but it suggests that the truth of instrumental reason is whatever knowledge about the external world – whether of
things, circumstances, consequences, or other persons – helps a person
to reach their goal in some action. If, then, someone has moral or non-
pragmatic concerns, they are understood to be entirely separate from
the exercise of instrumental reason. Now, we might ask: is this really

truth (i.e., is there a discrepancy between divinely given truth and the truth implied in instrumental reason), and is such an imagined, limited power really a kind of reason? I'll say more about that below.

Instrumental reasoning is the ideal mental or logical process of attaining such facts and determinations. In some sense, a person engaged in instrumental reasoning is trying to come to a "correct" decision; they are focused on particular, yet assumed rules of logic. There are very many ways in which the concept and related processes are analyzed by philosophers. As a broadly applicable definition that attempts to sidestep most of the disagreements, we can say that instrumental reasoning is the mental activity involved in correct evaluation, selection, and intention of suitable means for the given end(s).[125] For different persons, the word "suitable" can be taken as a placeholder for a variety of possible evaluative terms that relate to utility, such as efficiency or effectiveness. The word "correct" in this definition may be paired with the requirement of a maximal, adequate, sufficient, or long-term attainment of the end.

Given this definition of instrumental reasoning, some assumptions are implied in the common usage of the term, although they are usually not expressed. For example, there is an assumption of a real and generally desirable (independent of any additional and codetermining effects) relationship of causation between the means and the end; the means are actions or tools that cause the end in some way.[126] What is

---

[125] For a similar attempt at a broadly applicable definition, see Niko Kolodny and John Brunero, *The Stanford Encyclopedia of Philosophy* (Summer 2023 Edition), ed. Edward N. Zalta and Uri Nodelman, S.V. "Instrumental Rationality," https://plato.stanford.edu/archives/sum2023/entries/rationality-instrumental.

[126] Georg Spielthenner argues that the relation can be one of identity instead of causation. Even a relation of identity, however, entails using the presence of the means to attain the end, which seems to me to be a form of causation; the instrumentally rational person does not simply attain the end without the mediating

"suitable" therefore seems to exclude any evaluative criteria that is perceived as exogenous to (not strictly or necessarily associated with) the causal relationship – expressed by terms like moral, good, dignified, beautiful, etc.

Another assumption is that it is indeed possible to measure and evaluate the "correctness" of one's identification and utilization of available means. Although oriented to a practical end, this is essentially a process of discerning factual truth. A correct use of instrumental reasoning is based on the internal logical structure of the mental process and on the factual perception of means and consequences, but not on its proper place within a supreme goal of happiness through a personal relation with the Divine, let alone its relation to higher principles of morality, beauty, and integrity. A third assumption is that the end is desirable for some reason that is independent of whatever means are selected in attaining it (even though, in practice, the reasoning person may not be fully able to articulate that reason).

The combination of these assumptions appears to be intended to protect the reasoning model from the trap of interminably altering ends to suit the available means and vice versa.[127] Consider, for example, what happens if a person desires to cross a busy city street. If the end of getting to the other side of the street were not independently desirable, alternate choices of means could affect the desirability of the end, or even supersede it in importance. The person could end up in the middle of the street, find that they really enjoy the game of moving through the chaotic crowd while the light is green – or they might enjoy the view of some famous landmark several blocks away – and decide to remain on

---

means. See Georg Spielthenner, "Instrumental Reasoning Reconsidered," *European Journal of Analytic Philosophy* 4, no.1 (2008), 59-76.

[127] Christopher M. Reilly, "How Artificial Intelligence Technology Encourages the Vice of Acedia," *Divus Thomas* 126, no. 2 (2023), 154-5.

the street, bobbing and weaving among the hurried pedestrians even as the light turns red. This would not be instrumentally rational behavior (nor would it be safe)! Also, if that person could not be certain that it is possible to accurately measure and evaluate the available means for the given end of getting to the other side of the street, such as weaving among the people or stepping further to the side to avoid the heaviest rush of the crowd, the difficulty of the choice may lead to abandoning or curtailing the requirements of the end, or it may result in the model bringing in other criteria (such as other ends or non-instrumental values like justice and charity) to prefer alternative means that are not, in fact, the most efficient or effective in attaining the original end. They might stand paralyzed on the curb, afraid of trying to cross the street, or they might decide that it is either emotionally distasteful to be stuck in the middle of such a crowd or uncharitable to push their way through.

When the reasoning subject is focused primarily on a useful end, then confidence, speed, efficiency, and stability within the reasoning process itself – that is, the selection of means – can appear to enhance such an objective. If the street-crosser has a goal of getting to an urgent appointment, they may be more willing to push through the crowd in a straight line or gently shove some weaker persons out of the way. Some ends, however, are derived from more fundamental desires, whether through deductive reasoning or some similarity or analogy between the desires for the basic and derived ends.[128] Crossing the street, for example, may be associated at times with the more fundamental goal of reaching safety. When the derivation of such an end is influenced by circumstances that are in turn affected by the choice of means, it can be quite difficult to simultaneously identify appropriately derived ends as

---

[128] Steven Ellis, "The Varieties of Instrumental Rationality," *The Southern Journal of Philosophy* 46 (2008), 204.

well as the suitable means, and various mental shortcuts may be used to conveniently conclude the mental process. The street-crosser seeking safety will be confronted with different choices in crossing the street that, in turn, affect their safety (such as walking near other people during the Covid pandemic or stepping onto a manhole in the street). It may be impossible to avoid strategies like applying non-instrumental values as exclusionary principles or as ranking criteria of either the end or means.[129] Subjective beliefs or ideologies regarding causality, the relevance of circumstances and consequences, and the measurement of criteria like efficiency and utility can significantly influence the reasoning process in ways that may be unconscious or unintended by the subject.

In common parlance, it seems that most usage of the terms instrumental *reason* or instrumental *rationality* refers, in fact, to this *reasoning* process. Consideration of instrumental reasoning raises questions like: How can the subject most suitably cause the end? How can the subject most suitably evaluate alternative means? Is (and how is) the process of instrumental reasoning intertwined with interests, desires, emotions, perceptions, etc.? Can instrumental reasoning be evaluated morally, and what are the moral implications of its application and conclusions?

Instrumental rationality is a persistent disposition (a habitual way of thinking and acting) by which persons emphasize instrumental reasoning over other ways of thinking and behaving. From a Christian theological perspective, the following section will review many expressions

---

[129] See Christine Korsgaard, "The Normativity of Instrumental Reason," in *Ethics and Practical Reason*, ed. Garrett Cullity and Berys Gaut, (Oxford: Clarendon Press, 1997); Warren Quinn, "Putting Rationality in its Place," in *Morality and Action* (Cambridge: Cambridge University Press, 1993), 26-50; and Amartya Sen, *Rationality and Freedom* (Cambridge: Harvard University Press, 2002), 314.

of concern about instrumental reasoning and rationality, at least when applied to the higher, spiritual goods of human life, moral discernment, our relationship with God, and virtue or holiness in general. From the perspective of common sense, however, it at least appears that we cannot do without frequent attempts at instrumental reasoning about practical affairs. We are bodily, limited creatures who also wrestle with cognitive limits on our intellectual powers, and therefore we do not have the capacity for simply willing our final end – or even our relatively higher ends such as virtuous living – into being, but must work with a contextually restricted availability of the limited physical and strategic means that imperfectly, contingently, and variously further our ends. Each person's relationship with their environment, other people, and even their experienced self is, in many ways, the relation of subject to object, craftsman to tool, or user to instrument.

On the other hand, the principal end of human beings is a loving relationship with God, something that cannot be attained through instrumental reasoning and its associated action because love is gratuitous, self-giving, and mutually affirming (rather than oriented to alteration, manipulation, strategic use, or acquisition of an object by the opposed subject), and because a controlling stance toward our omnipotent God is intrinsically self-defeating and unfulfilling; we can't *make* God love and embrace us no matter what means and strategies we apply. Consider Pope Benedict XVI's teaching about love in the dual forms of eros and agape:

> Even if eros is at first mainly covetous and ascending, a fascination for the great promise of happiness, in drawing near to the other, it is less and less concerned with itself, increasingly seeks the happiness of the other, is concerned more and more with the beloved, bestows

itself and wants to "be there for" the other. The element of agape thus enters into this love, for otherwise eros is impoverished and even loses its own nature. On the other hand, man cannot live by oblative, descending love alone. He cannot always give, he must also receive. Anyone who wishes to give love must also receive love as a gift. [130]

If, as St. Thomas Aquinas taught, willing the means is really a matter of willing the end, [131] then we might at least consider the possibility that any strategic action not ultimately oriented to the final end of a non-instrumental, loving relationship with God would seem to be essentially immoral – a distraction or divergence from the principal end of the person. As will be discussed below, instrumental reasoning in any instance is only a partial implementation of a more holistic reason that finds truth in Divine love. This is why a moral evaluation of the "correctness" of instrumental reasoning or "suitability" of choosing means toward ends always seems to miss the point of true human fulfillment, and why the notion of a strictly defined power of instrumental *reason* may in fact be an artificial and nonsensical division of the human pursuit of truth.

In this work, neither the absolute immorality of instrumental reasoning in itself, nor the falsity of instrumental reason, will be argued. The focus here is on instrumental *rationality*, and a moral judgment of all activities governed by instrumental reasoning is not necessary to show a connection between instrumental rationality, acedia, and AI. In the next sections, however, the grave concerns outlined above will be reviewed in the writings and communications of the Catholic popes. A

---

[130] Benedict XVI, *Deus caritas est* (Dec. 25, 2005), 7.

[131] Aquinas, *Summa Theologica* I-II, 8, 2; and 10, 2, ad 3.

deep ambivalence about instrumental reasoning and rationality seems to be warranted.

The definition given above for instrumental rationality can be specified further by describing it as a persistent disposition by which a person emphasizes their successful (e.g. effective, efficient, maximal) attainment (e.g. acquisition, possession, control, use) of intermediate goods and means that further one or more given ends – rather than attending to the choice and guidance of appropriate (e.g. good, moral, fulfilling) ends. The definition is rather broad and flexible because it attempts to describe a disposition, or a category of dispositions, that is, in fact, present in the intentional actions of real persons, and this disposition is not lived or experienced in a consistently distinct manner – not even in the awareness of these subjects. Unlike instrumental *reason* or instrumental *reasoning*, this is a descriptive term based on human action and intentions, not an abstract concept that can be defined in a purely speculative exercise. As such, it also implicates the subject's will and is therefore intrinsically a morally relevant practice.[132] We might say loosely that a person exhibiting instrumental rationality is regularly attempting to engage in instrumental reasoning, with the caveat that such reasoning may be an ideal that is, in practice, undefinable and logically confused (as explained briefly above). This is why the term instrumental *rationality*, with its focus on actual dispositions of real subjects, is preferable for our analysis here of acedia and the motivational relationship of AI to sinful behavior. The primary feature is its practical and procedural focus on means rather than ends, oriented to action.

The definition of instrumental rationality proposed here does not match exactly, but is derived from, the concept found in the sociological theorist Max Weber's writings. One way in which Weber presents

---

[132] Aquinas, *Summa Theologica* I-II, 18, 4; John Paul II, *Veritatis splendor* (Aug. 6, 1993), 78.

instrumental rationality, or *zweckrational*, is in contrast to values rationality. He describes the two opposed types of rationality rather succinctly:

> (1) instrumentally rational, that is, determined by expectations as to the behavior of objects in the environment and of other human beings; these expectations are used as "conditions" or "means" for the attainment of the actor's own rationally pursued and calculated ends; (2) value-rational, that is, determined by a conscious belief in the value for its own sake of some ethical, aesthetic, religious, or other form of behavior, independently of its prospects of success.[133]

According to Weber, if action is focused on one or more absolute values, it will be seen as irrational from the instrumental perspective, because it will be focused less on consequences. While this description and simple opposition between instrumental and values rationalities has been very influential, it is too far from the definition of instrumental rationality that will suit our purposes in this work, not least because Weber is concerned with calculated "expectations" of social behavior rather than the broader motive of utility that often drives evaluation of material, human, or social means, especially in the development and application of modern technology.[134] Also, Weber perceives a zero-sum, winner-takes-all competition for the attention of acting subjects in which greater concern for values necessarily undermines concern for preferred consequences. He does not consider here that the concern for

---

[133] Max Weber, *Economy and Society*, ed. Guenther Roth and Claus Wittich (Berkeley: University of California Press, 1978), 25–6.

[134] Weber, *Economy and Society*, 4.

certain values, such as the virtues of perseverance, temperance, and for-titude, can easily enhance the vigor with which a person engages in in-strumentally productive activity. The proposed contest between instru-mental and values rationality in *Economy and Society* is therefore too elegant a schema for real-world application and understanding.

If we examine the corpus of Weber's works, however, a different, more complex theory emerges. The discussion of instrumental ration-ality appears in several of Weber's works with somewhat different theo-retical contexts and vocabulary; the following summary draws on Ste-phen Kalberg's schema of Weber's theory of instrumental rationality, chosen for its accuracy as well as its easy applicability to our analysis of acedia and AI in this work.[135] Weber focuses on the study of social ac-tion – behavior toward and among persons – which he divides rather arbitrarily between the categories of affectual, traditional, value-ra-tional, and means-end rational.[136] The category of means-end rational action is easily confused with the common interpretation of instrumen-tal *reasoning*, but it is a category of social action, not of thought alone, nor is it so broad that it captures action oriented to non-personal ob-jects.[137] Perhaps the best way to characterize it is as bearing a distinct kind of motivation: to attain one's social ends through judicious use of available means, or to calculate useful ends according to available means. It is theoretically and experientially distinct from social action motivated by emotions (affectual social action), loyalty and consistency

---

[135] Stephen Kalberg "Max Weber's Types of Rationality: Cornerstones for the Analysis of Rationalization Processes in History," *The American Journal of Sociol-ogy*, 85, no. 5 (1980), 1145-1179.

[136] Weber, *Economy and Society*, 63-110.

[137] "[T]he end, the means, and the secondary results are rationally taken into account and weighed," Weber, *Economy and Society*, 26.

(traditional social action), or values and principles (value-rational social action).[138]

Characterizing the means-end rational and value-rational types of action, Weber describes four types of rationalities that represent diverse attempts to order thoughts and social relations in such a way that regularity, meaningfulness, and unity can be discerned in the face of the unpredictability of real life.[139] Practical rationality appears in means-end rational action, is motivated by individual interests, and involves selection and calculation of the best ends for attaining those interests.[140] Formal rationality, most commonly featured in bureaucratic organizations, is also found in means-end rational action but is governed by the application of rules to all (at least within a particular category or group).[141] Theoretical (or intellectual) rationality is focused on both holistic and precise explanation of reality through concepts, logical and causal reasoning, and symbolic language (i.e., mathematics and formal logic).[142] Substantive rationality stands out from all the others as an

---

[138] Weber, *Economy and Society*, 63-110.

[139] Kalberg draws these types of rationality from a variety of Max Weber's writings. He lists: "The Social Psychology of the World Religions" in *From Max Weber: Essays in Sociology*, ed. and trans. Hans H. Gerth and C. Wright Mills, (New York: Oxford University Press, 1958), 293-94; *The Protestant Ethic and the Spirit of Capitalism*, trans. Talcott Parsons, (New York: Scribner's, 1958), 26, 77-78; *Economy and Society*, 30, 85, 424, 809, 333; *The Religion of China*, ed. and trans. Hans H. Gerth, (New York: Free Press, 1951), 226; and *Ancient Judaism*, ed. and trans. Hans H. Gerth and Don Martindale, (New York: Free Press, 1952), 425-26, n. 1.

[140] Weber, *The Protestant Ethic*, 77; Weber, "The Social Psychology of the World Religions," 293; Kalberg, "Max Weber's Types of Rationality," 1151-2.

[141] Weber, *Economy and Society*, 975, 226; Weber, "The Social Psychology of the World Religions," 295; Kalberg, "Max Weber's Types of Rationality," 1158-9.

[142] Weber, *Economy and Society*, 85-86; Weber, "The Social Psychology of the World Religions," 294; Kalberg, "Max Weber's Types of Rationality," 1152-5.

accompaniment to values-rational action that chooses and judges actions according to a variously integrated set of values – a value postulate.[143]

There is much in Weber's theories, philosophical perspective, and historical interpretations that is worth considering or debating, but our purpose here is only to draw on his schema of action and rationality to develop a working definition of instrumental rationality in support of a theological argument. His approach to rationality as a characteristic of action, rather than as an ideal process abstracted from real conditions and subjects, is useful in that it allows us to sidestep the many difficulties found in the concept of instrumental *reasoning* as a mental process and to implicate the human will in the attempt at such rationality, thereby creating an opening for moral evaluation of the disposition and its manifestations. Weber is, however, concerned only with social action, while our concept will apply more broadly to human action in general. Weber also usefully recognizes that there is potential for one type of rationality to interfere or dominate action governed by another type of rationality, as he emphasizes in regard to the rise of powerful bureaucracies that crowd out values-rational action with frequently expanding imposition of formally rational social relations and supportive material structures (e.g. pay scales, procedures oriented around paper forms, military organization, etc.).[144]

In the definition proposed above for instrumental rationality, the various alternatives for "successful attainment" of means and goods, contrasted with the alternative terms for concern about "appropriate" ends, give due consideration to Weber's four types of rationality within

---

[143] Weber, "The Social Psychology of the World Religions," 293; Kalberg, "Max Weber's Types of Rationality," 1155-7.

[144] Weber, *Economy and Society*, 26, 85-6, 226, 975; "The Social Psychology of the World Religions," 295.

a single definition. Our definition of instrumental rationality excludes substantive rationality because the absolute criteria of substantive evaluation of the rationality of action is, or is at least perceived to be, exogenous to the presumed causal relationship between the means and end. We can characterize instrumental rationality as the disposition associated with means-end action, in particular the application of formal and practical types of rationality (in Weber's terms) as well as theoretical abstraction from reality in so far as it supports the formal and practical efforts at control, predictability, and regularity in the world.

Another important concept for this work is Weber's notion of rationalization, a process by which a society or social unit adopts regular patterns of action that favor one or more of the types of rationality described above; in a fully rationalized society, "one can, in principle, master all things by calculation."[145] The essence of rationalization is the manifestation of stable, relatively inflexible patterns in the real behavior of persons; it is the triumph of certain motivations and associated thought processes over social reality through time. The social patterns and structures, including material incentives and sanctions, may reinforce those motivations and thought patterns so that the instituted type of rationality progressively dominates additional actions and spheres of social life. Rationalization is therefore dynamic and often – although not necessarily – self-reinforcing and expanding. It also manifests differently according to historical or cultural conditions and the associated commitments to particular interests, ideas, and values. There is no inevitable progression, even in the most rationalized cultures, and different processes of rationalization can simultaneously occur in separate spheres of life – although one kind of rationality may come to dominate

---

[145] Max Weber, "Science as a Vocation," in *From Max Weber: Essays in Sociology*, ed. H.H. Gerth and C. Wright Mills, (Oxford: Oxford University Press, 1946), 139; Kalberg, "Max Weber's Types of Rationality," 1170-1.

or guide the relative influence of other types. For the purposes of this work, instrumental rationalization is a process whereby instrumental rationality gains intellectual and functional status as a normative canon of human action in at least some spheres or distinct activities of life, if not an entire way of life pervading most human activity. Moreover, we will place greater emphasis than Weber does on the material structures, incentives, and constraints that reinforce rationalization of instrumental rationality in contemporary Western culture.

A Path to Acedia

I have referred to instrumental rationality as a vice. For St. Thomas Aquinas, who is frequently cited for his wisdom on this topic, "the vice of any thing consists in its being disposed in a manner not befitting its nature," meaning that a vice in a person is some habitual quality that is contrary to what that person naturally needs and desires.[146] But what is natural? Aristotle showed that this is most properly the essential characteristics of the species (humanity) of which the person is part of; and the core essential characteristic of humanity – that which distinguishes it from and above other species of animals – is its power of reason. A personal vice, opposed to the person's nature, is therefore a habit that is irrational, given the full nature, true needs, and ultimate desires of the person. The worst vice would be a habit of behavior or thought that turns a person away from their beatitude, their real happiness, which is eternal life and friendship with God, for to willingly spurn or invalidate one's beatitude would be the height of irrationality.

Instrumental rationality is a vice because it is a habitual disposition that makes striving for one's beatitude difficult, even frustrating, and

---

[146] Thomas Aquinas, *Summa Theologica* I-II, q.71, a.2.

sometimes to an extreme degree. It is fundamentally not possible to reliably secure eternal life with God through some means that cause the end to be attained, even though our efforts in attaining holiness and virtue, and in accomplishing good works, help to form us into better Christians – better participants in our relationships with God. This is first of all because the most important virtues, the theological virtues of faith, hope, and love, cannot be acquired through human effort but rely on God's infusion of grace for our enjoyment of them. The human intellect and will are limited powers; we cannot independently, without the gifts of the Holy Spirit, reach understanding of our uncreated God nor willfully desire friendship with God enough to establish a true union or eternal friendship. Also, because God is omnipotent, unchanging, and perfect, we know that there is no possibility of instrumentally manipulating either the created world or God's will in order to acquire God's friendship.

Even the nature of love and friendship is manifested in decidedly non-instrumental relationships between beings, for love and friendship are not primarily oriented to acquiring, controlling, achieving, or maximizing. As I have written previously,

> [L]ove – even between human persons – is not a state of being that is controlled or independently willed, but a mutual act of gratuitous self-giving. The meanings and valuations discovered or intended by the participants in the variety of acts comprising a loving relationship are ultimately subjective, contingent, and often negotiated, so no participant can have consistently reliable instrumental control of the relationship through their individual choices. Such attempts at causal manipulation, in fact, have the tendency to generate power-based

opposition, hurt, mistrust, and indignity that subvert the end of a sustained loving connection.[147]

Such love is transformative. We don't simply reach out and interact with the loved object or person as if there were an invisible line of separation that we dare not cross. In love, we *are moved* by what we love, our hearts *melt*, and we personally feel the experiences of the loved object or person with empathy as we willingly transform ourselves into partners who are receptive and, in some sense, able to unite with the loved object. This is especially true of our relationship with God, who is not just a created being but is Being itself, of an infinitely greater order of being than we can hope to attain, yet so loving and self-giving that he extends the grace for us to transform in holiness and strive to be nearer to him. To try to simply acquire or control that personal end of beatitude would not only be futile, it would drive the person ever further from the theological virtues that are not merely means but manifestations of the limited perfection of the person in their relationship with God.

Another reason that instrumental rationality is a vice is because it leads a person toward means and ends that have limited potential for enabling true joy in the person, or it distracts from the more fundamental ends related to one's beatitude. When a person is primarily focused on such concerns as urgency, ease or efficiency, simplicity, and control or acquisition of perceived goods, they are not focused on the exercise of reason for attaining beatitude. For example, in order to feasibly choose particular means for a given end, a person will likely want to prioritize only one or a few competing criteria and methods of evaluation. If our end is limited and focused on acquiring some goods or

---

[147] Christopher M. Reilly, "How Artificial Intelligence Technology Encourages the Vice of Acedia," *Divus Thomas* 126, no.2 (2023), 156; Francis, *Laudato si'*, 105.

achieving some goal, we are strongly tempted to give priority to such criteria as efficiency and productivity over the aesthetic appeal, moral value, or self-transformative opportunity associated with alternative means. Even the cognitive processes involved in evaluating efficiency and morality or beauty are very different, making it unlikely that we will engage in such practices as contemplation or moral discernment when a current task calls for urgency or achievement. We find ourselves habitually focused on these pragmatic criteria rather than what will bring us closer to God.

The consequence of instrumental rationality as a vice is dissatisfaction and frustration on the path to beatitude. If instrumental rationality is the dominant mode of being for a person, they will feel alienated from an end – their true beatitude – that seems ever more difficult to understand and experience. It will increasingly appear that the person must struggle in a futile effort to attain the joy of a relationship with God, and the instrumentally oriented person will become even more discouraged. The result is the manifestation of acedia as sorrow, anxiety, and resigned or recalcitrant sloth.

Three Popes' Wisdom

Pope Saint John Paul II:

For Pope St. John Paul II, instrumental rationality is a problem for persons' striving toward truth, real freedom, and meaning. Late modern philosophy is therefore deficient in its loss of the sense of a holistic rationality open to faith and revelation.

> Other forms of rationality have acquired an ever higher profile, making philosophical learning appear all the more peripheral. These forms of rationality are

directed not towards the contemplation of truth and the search for the ultimate goal and meaning of life; but instead, as "instrumental reason", they are directed—actually or potentially—towards the promotion of utilitarian ends, towards enjoyment or power.[148]

Here, although St. John Paul uses the term instrumental *reason*, he also identifies it as a kind of *rationality*, one which is associated with a disposition toward action that furthers the "utilitarian ends" of enjoyment and power. This instrumental rationality is therefore closely associated with particular ends – it is not a merely neutral process in regard to ethical purposes – and involves a kind of thinking process that emphasizes calculation of means to promote the material welfare of individuals. St. John Paul also declares that such philosophers "have abandoned the search for truth in itself and made their sole aim the attainment of a subjective certainty or a pragmatic sense of utility. This in turn has obscured the true dignity of reason, which is no longer equipped to know the truth and to seek the absolute."[149]

Our goal, according to St. John Paul, is to orient our reasoning activity toward truth, rather than toward strategic exercise of power and attainment of utility through a narrow kind of individual freedom. "Authentic freedom is never freedom 'from' the truth but always freedom 'in' the truth."[150] The modern distortion falsely offers hope that "people achieve such dignity when they free themselves from all subservience

---

[148] John Paul II, *Fides et ratio* (Sep. 14, 1998), 47.

[149] John Paul II, *Fides et ratio*, 47.

[150] John Paul II, *Veritatis splendor*, "The Splendor of Truth" (August 6, 1993), 64.

to their feelings, and in a free choice of the good, pursue their own end by effectively and assiduously marshalling the appropriate means."[151]

> [R]eason, in its one-sided concern to investigate human subjectivity, seems to have forgotten that men and women are always called to direct their steps towards a truth which transcends them. Sundered from that truth, individuals are at the mercy of caprice, and their state as person ends up being judged by pragmatic criteria based essentially upon experimental data, in the mistaken belief that technology must dominate all. ... Abandoning the investigation of being, modern philosophical research has concentrated instead upon human knowing. Rather than make use of the human capacity to know the truth, modern philosophy has preferred to accentuate the ways in which this capacity is limited and conditioned.[152]

There is a path from the narrow implementation of reason to pragmatism and even utilitarianism. With a restricted focus on generating useful knowledge from the results of experimental science, philosophers have lost sight of "being" itself with a skeptical emphasis on the limits of human perception and other cognitive powers. This, in turn, leads to domination – of the material world through technology and also of the social world. Utilitarianism as an ethical formula is the result: "[The reliance on scientific verification], as we know, results in agnosticism in theory and utilitarianism in practice and in ethics. ... Utilitarianism is a civilization of production and of use, a civilization of 'things'

---

[151] John Paul II, *Veritatis splendor*, 21.
[152] John Paul II, *Fides et ratio*, 5.

and not of 'persons', a civilization in which persons are used in the same way as things are used."[153]

Such a reduced form of rationality does not serve the principal end of human persons, who are also spiritual beings. "People seek an absolute which might give to all their searching a meaning and an answer—something ultimate, which might serve as the ground of all things. In other words, they seek a final explanation, a supreme value, which refers to nothing beyond itself and which puts an end to all questioning."[154] This final explanation is truth. Once human beings recognize the centrality of truth to the meaning of their action, individual freedom, power, and certainty are no longer perceived as if they are opposed to the "yoke" of divine law.

> The rightful autonomy of the practical reason means that man possesses in himself his own law, received from the Creator. Nevertheless, the autonomy of reason cannot mean that reason itself creates values and moral norms. Were this autonomy to imply a denial of the participation of the practical reason in the wisdom of the divine Creator and Lawgiver, or were it to suggest a freedom which creates moral norms, on the basis of historical contingencies or the diversity of societies and cultures, this sort of alleged autonomy would contradict the Church's teaching on the truth about man. It would be the death of true freedom: "But of the tree of the knowledge of good and evil you shall not eat, for in the day that you eat of it you shall die" (Gen 2:17).[155]

---

[153] John Paul II, *Evangelium vitae* (March 25, 1995), 13.

[154] John Paul II, *Fides et ratio*, 27.

[155] John Paul II, *Veritatis splendor*, 40.

There is an existential aspect of the search for truth in which human action strives toward the higher moral values. "The acting subject personally assimilates the truth contained in the law. He appropriates this truth of his being and makes it his own by his acts and the corresponding virtues."[156] In his philosophical work *Person and Act*, written before becoming pope, St. John Paul (Karol Wojtyła) demonstrates the crucial ethical nature of the human act, which implements free choice of values and, by accessing the higher values, thereby achieves transcendence of one's individual being; a person who chooses the good becomes a morally good person in a higher sense.[157] That philosophy is later reflected in St. John Paul's papal documents: "The truth of these values is to be found not by turning in on oneself but by opening oneself to apprehend that truth even at levels which transcend the person."[158]

This transcendence is not an exercise of pure human will, however, for it is one's relationship with God and enjoyment of divine grace that enables and guides the transcendent act. "It is important to realize that among the many questions surfacing in your minds, the decisive ones are not about 'what.' The basic question is 'who': to whom am I to go? whom am I to follow? to whom should I entrust my life?"[159] On the other hand, the primary consequence of a narrowing of rationality and reliance on technology is a reduction of love. "Everything contrary to the civilization of love is contrary to the whole truth about man and becomes a threat to him … And what is this danger? It is the loss of the

---

[156] John Paul II, *Veritatis splendor*, 52.

[157] Karol Wojtyła, *Person and Act and Related Essays*, Vol. 1, trans. Grzegorz Ignatik (Washington, D.C.: The Catholic University of America Press, 2021), 207-91.

[158] John Paul II, *Fides et ratio*, 25.

[159] John Paul II, homily, closing of World Youth Day (Aug. 20, 2000), 3.

truth about one's own self and about the family, together with the risk of a loss of freedom and consequently of a loss of love itself."[160] Moreover, the narrowing of rationality and loss of truth undermines a person's relationship with God; the central event of that relationship becomes unintelligible from an instrumentally rational perspective.

> The crucified Son of God is the historic event upon which every attempt of the mind to construct an adequate explanation of the meaning of existence upon merely human argumentation comes to grief. The true key-point, which challenges every philosophy, is Jesus Christ's death on the Cross. It is here that every attempt to reduce the Father's saving plan to purely human logic is doomed to failure. … Reason cannot eliminate the mystery of love which the Cross represents, while the Cross can give to reason the ultimate answer which it seeks. It is not the wisdom of words, but the Word of Wisdom which Saint Paul offers as the criterion of both truth and salvation.[161]

While instrumental rationality privileges the empirical evidence that informs useful, means-oriented action and its discovery through experimental science, a holistic rationality integrates the evidence, relation, and disposition of faith. "Christian Revelation is the true lodestar of men and women as they strive to make their way amid the pressures of an immanentist habit of mind and the constrictions of a technocratic

---

[160] John Paul II, *Gratissimam sane* "Letter to Families" (Feb. 2, 1994), 13.
[161] John Paul II, *Fides et ratio*, 23.

logic."[162] Striving for truth both requires and compels an integration of faith and reason.

> [R]eason has its own specific field in which it can enquire and understand, restricted only by its finiteness before the infinite mystery of God. Revelation therefore introduces into our history a universal and ultimate truth which stirs the human mind to ceaseless effort; indeed, it impels reason continually to extend the range of its knowledge until it senses that it has done all in its power, leaving no stone unturned.[163]

Faith is not merely a belief in a form of non-empirical evidence that is inserted into the calculations of reason, but it is a willed choice to orient reason toward what is good and true. "The fool thinks that he knows many things, but really he is incapable of fixing his gaze on the things that truly matter. Therefore he can neither order his mind (Prov 1:7) nor assume a correct attitude to himself or to the world around him."[164]

For St. John Paul, humanity's obsessive attitude toward modern technology is a great problem. "According to the Enlightenment mentality, the world does not need God's love. The world is self-sufficient. And God, in turn, is not, above all, love. If anything, he is intellect, an intellect that eternally knows. … the demigod of modern technology. This is the world that must make man happy."[165] Such a world, however,

---

[162] John Paul II, *Fides et ratio*, 15.

[163] John Paul II, *Fides et ratio*, 14.

[164] John Paul II, *Fides et ratio*, 18.

[165] John Paul II, *Crossing the Threshold of Hope*, ed. Vittorio Messori (New York: Alfred A. Knopf, 2005), 55.

does not lead to true happiness. What is required is nothing less than a reversal of deeply ingrained priorities of the will, empowered by God's grace. "The essential meaning of this 'kingship' and 'dominion' of man over the visible world, which the Creator himself gave man for his task, consists in the priority of ethics over technology, in the primacy of the person over things, and in the superiority of spirit over matter."[166]

Another mode of human activity in which the problem of instrumental rationality becomes apparent is in productive work that often results in "alienation."

> The man of today seems ever to be under threat from what he produces, that is to say from the result of the work of his hands and, even more so, of the work of his intellect and the tendencies of his will. All too soon, and often in an unforeseeable way, what this manifold activity of man yields is not only subject to 'alienation', in the sense that it is simply taken away from the person who produces it, but rather it turns against man himself, at least in part, through the indirect consequences of its effects returning on himself. … Man therefore lives increasingly in fear.[167]

Alienation results from a "reversal of means and ends;" calculating and marshalling the means for efficient and maximal production leads to treatment of the subjects of work as mere means.[168] The true end of the worker is lost in the activity, often due to a power imbalance with the employers, but a person is called to do much more through their

---

[166] John Paul II, *Redemptor hominis* (March 4, 1979), 49.

[167] John Paul II, *Redemptor hominis*, 53.

[168] John Paul II, *Centesimus annus* (May 1, 1991), 41.

work. "A man is alienated if he refuses to transcend himself and to live the experience of self-giving and of the formation of an authentic human community oriented towards his final destiny, which is God."[169]

In the considerations associated with bioethics – especially including beginning-of-life problems of abortion, contraception, in vitro fertilization (IVF) as well as end-of-life problems of euthanasia and physician-assisted suicide – instrumental rationality is an underlying theme that is rarely addressed explicitly yet heavily influences moral thinking. St. John Paul's writings and teachings referred to as the "theology of the body" can be characterized as a great effort to warn against the selfish preference for instrumental rationality when dominating or using new human beings to achieve preferred acquisition (or prevention) of children. Contraception, for example, is described as a pursuit of utility that undermines the sacred relationality between spouses, with their child, and with God; such practices as contraception and abortion "imply a self-centered concept of freedom, which regards procreation as an obstacle to personal fulfilment."[170] IVF and surrogate motherhood are narrowly rational techniques to gain from the efforts of others in a highly undignified transfer of children as property.[171] The core value in the theology of the body, on the other hand, is Trinitarian relationality (the kind of love and complementarity found between the persons of the Holy Trinity) that suffuses the couple's participation in the divine gift of human life – a relationality that is self-giving, celebratory, reverent, and above all loving.[172] We can see that such relationality is diametrically

---

[169] John Paul II, *Centesimus annus*, 41.

[170] John Paul II, *Evangelium vitae*, 13.

[171] John Paul II, *Evangelium vitae*, 63.

[172] Congregation for the Doctrine of the Faith, *Instruction on Respect for Human Life in Its Origin and on the Dignity of Procreation: Replies to Certain Questions of the Day* "Donum Vitae" (Feb. 22, 1987), II, A, 3.

opposed to the logic of instrumental rationality; attitudes of reverence and joy are considered to be irrelevant to the calculations of instrumental reason, and the activity of self-giving through a loving, spiritual union with the beloved object is flatly irrational from the perspective of the narrower rationality. St. John Paul writes:

> In its most profound reality, love is essentially a gift; and conjugal love, while leading the spouses to the reciprocal "knowledge," which makes them "one flesh," does not end with the couple, because it makes them capable of the greatest possible gift, the gift by which they become cooperators with God for giving life to a new human person.[173]

In this passage, there is no violent, interest-led, or acquisitive opposition between subject and object, attained through efficient use of means, but the overarching celebration of unity in purpose and sharing. The very nature, meaning, and attractiveness of the means comes solely from the holiness of the end. Technical or technological means which undermine the unitive meaning of the marital act, whether contraception or the killing of conceived children in abortion and IVF (killed or frozen), are not simply prohibited and repugnant to the dignity of the child and spouses, but sever the active participants from their naturally principal good, which is a right relationship with God.[174] Instrumental rationality in such cases is truly the abrogation of right reason.

In St. John Paul's theology of the body, he recognizes the human person as a teleological (essentially purposeful) being with an inherent and principal good found in relationship with God. It is therefore good

---

[173] John Paul II, *Familiaris Consortio* (Nov. 22, 1981), 14.
[174] John Paul II, *Evangelium vitae*, 23.

to utilize available means to grow in self-mastery, for this is an extension of one's natural teleology. Instrumental rationality, on the other hand, results from a strategic and existential opposition between the person as subject and the external world as object. This radical liberty found in Descartes' purely spiritual agent, for whom "freewill is in itself the noblest thing we can have because it makes us in a certain manner equal to God and exempts us from being his subjects," is hardly compatible with St. John Paul's understanding of self-mastery and freedom found in love.[175] The pursuit of human power led Descartes' and Francis Bacon's to an insistence on a "technological" rationality that would make such power effective.[176] For St. John Paul, however, this preference for mechanical means serves only the limited purposes and ontology of instrumental rationality:

> This extension of the sphere of the means of "the domination ... of the forces of nature" menaces the human person for whom the method of "self-mastery" is and remain specific. The mastery of self corresponds to the fundamental constitution of the person; it is indeed a "natural" method. On the contrary, the resort to artificial means destroys the constitutive dimension of the person. It deprives man of the subjectivity proper to him and makes him an object of manipulation.[177]

---

[175] René Descartes, *Philosophical Letters*, trans. Anthony Kenny (Minneapolis: University of Minnesota, 1970), 228.

[176] Michael Waldstein, "Introduction," in *Man and Woman He Created Them: A Theology of the Body*, trans. Michael Waldstein (Boston: Pauline Books & Media, 2006), 36-44.

[177] John Paul II, *The Theology of the Body: Human Love in the Divine Plan* (Boston: Pauline Books and Media, 1997), 397.

As technical options for manipulating human biology proliferated, the concept of technical domination became prominent during the pontificate of Pope St. John Paul II.[178] Pope Paul VI had previously stated that, in its teaching on manipulation of procreation, the Church "urges man not to betray his personal responsibilities by putting all his faith in technical expedients."[179] In his encyclical *Evangelium vitae*, St. John Paul reiterated and expanded this teaching in regard to the moral prohibition of direct abortion, physician-assisted suicide, in vitro fertilization (IVF), and eugenics-oriented genetic testing.[180] The Congregation for the Doctrine of the Faith issued the document *Donum Vitae* during St. John Paul's pontificate, declaring that:

> The one conceived must be the fruit of his parents' love. He cannot be desired or conceived as the product of an intervention of medical or biological techniques; that would be equivalent to reducing him to an object of scientific technology. No one may subject the coming of a child into the world to conditions of technical efficiency which are to be evaluated according to standards of control and dominion.[181]

Moreover, technical domination can take the form of mastery over the human body, including non-therapeutic genetic enhancement and

---

[178] For a fuller discussion, see Christopher M. Reilly, "Technological Domination: Its Moral Significance in Bioethics," *National Catholic Bioethics Quarterly* 23, no. 1 (Spring 2023), 23–35.

[179] Paul VI, *Humanae vitae* (July 25, 1968), 18.

[180] John Paul II, *Evangelium vitae*, 4, 14, 22, 64, and 81.

[181] Congregation for the Doctrine of the Faith, *Donum Vitae*, II, B, 4, c.

surrogacy techniques and services.[182] The ideology of instrumental rationality exacerbates problems with misuse of technologies: "One cannot derive criteria for guidance from mere technical efficiency, from research's possible usefulness to some at the expense of others, or, worse still, from prevailing ideologies."[183] St. John Paul warned that technical domination is, in itself, insidious and expands in both capacity and the reach of associated ideologies, for "the various techniques of artificial reproduction, which would seem to be at the service of life and which are frequently used with this intention, actually open the door to new threats against life."[184]

Pope Benedict XVI:

Pope Benedict XVI taught that truth can only be accessed by a holistic reason that reflects the ancient Greek concept of logos as "creative" reason. In contrast,

> [The] modern concept of reason is based, to put it briefly, on a synthesis between Platonism (Cartesianism) and empiricism, a synthesis confirmed by the success of technology. On the one hand it presupposes the mathematical structure of matter, its intrinsic rationality, which makes it possible to understand how matter works and use it efficiently: this basic premise is, so to speak, the Platonic element in the modern understanding of nature. On the other hand, there is nature's

---

[182] See Congregation for the Doctrine of the Faith, *Donum Vitae*, introduction, 4 and I, 6 and II, A, 2-3; John Paul II, *Address to the Thirty-Fifth General Assembly of the World Medical Association* (October 29, 1983).

[183] Congregation for the Doctrine of the Faith, *Donum Vitae*, introduction, 2.

[184] John Paul II, *Evangelium vitae*, 14.

capacity to be exploited for our purposes, and here only the possibility of verification or falsification through experimentation can yield decisive certainty.[185]

As Pope St. John Paul II also asserted, modern culture narrows the concept of reason to a calculative process that marshals empirical evidence for purely pragmatic, limited purposes of control and exploitation. "Self-communication," the self-giving of the subject in dialogue with real being, is cut off in an opposition of subject and object, on both epistemological and practical levels.

> Common to the whole Enlightenment is the will to emancipation, first in the sense of Kant's *sapere aude* — dare to use your reason for yourself. Kant is urging the individual reason to break free of the bonds of authority, which must all be subjected to critical scrutiny. Only what is accessible to the eyes of reason is allowed validity. This philosophical program is by its very nature a political one as well: reason shall reign, and in the end no other authority is admitted than that of reason. Only what is accessible to reason has validity; what is not reasonable, that is, not accessible to reason, cannot be binding either.[186]

Reason is therefore a logos that is much more that the restricted logic of instrumental rationality. To access God as truth, "We will

---

[185] Benedict XVI, "Faith, Reason and the University: Memories and Reflections," lecture (September 12, 2006), https://www.vatican.va/content/benedict-xvi/en/speeches/2006/september/documents/hf_ben-xvi_spe_20060912_university-regensburg.html.

[186] Joseph Ratzinger, "Truth and Freedom," *Communio*, 23 (1996), 20-21.

succeed in doing so only if reason and faith come together in a new way, if we overcome the self-imposed limitation of reason to the empirically falsifiable, and if we once more disclose its vast horizons."[187] Faith is the spur to engaging the totality of being, to "ask about the foundations of the totality of things."[188]

For Benedict, we are best with a "technological Prometheanism" that merges instrumental rationality, scientism, radical individualism, and radical freedom.[189] This is hardly the attitude that leads a person to their beatitude.

> From this union of the materialistic vision of man and the great development of technology a fundamentally atheist anthropology emerges. It presupposes that man is reduced to autonomous functions, the mind to the brain, human history to a destiny of self-realization. … The most dangerous snare of this current of thought is in fact the absolutization of man: man wants to be *absolutus*, freed from every bond and from every natural constitution. … This is a radical denial of the nature of

---

[187] Benedict XVI, "Faith, Reason and the University."

[188] Joseph Ratzinger, *The God of Jesus Christ: Meditations on the Triune God*, trans. Brian McNeil (San Francisco: Ignatius Press, 2008), 41. "Faith makes technical research and questioning possible, because it explains the rational character of the world and the orientation of the world toward man; but it is profoundly opposed to restricting thought exclusively to questions of function and usefulness. Faith challenges man to look beyond immediate usefulness and to ask about the foundations of the totality of things. Faith protects the contemplative and listening reason from attack by the merely instrumental reason."

[189] Benedict XVI, Address to Participants in the Plenary Meeting of the Pontifical Council "*Cor Unum*" (Jan. 19, 2013).

the creature and child in man, which ends in tragic loneliness.[190]

Instrumental rationality encourages a crisis in meaning, one which leaves radically "free," individual and isolated persons to become mere cogs in the wheel of technological progress. "To resist this eclipse of reason and to preserve its capacity for seeing the essential, for seeing God and man, for seeing what is good and what is true, is the common interest that must unite all people of good will. The very future of the world is at stake."[191]

According to Benedict, instrumental rationality and technological efforts are closely related to the dominance of a narrowly focused, modern science that inappropriately restricts persons' knowledge of objects – of a true relationship with reality – while its purpose is essentially tied to its technical utility that is, in turn, motivated by control and exercise of power. "But what is the basis of this new era? It is the new correlation of experiment and method that enables man to arrive at an interpretation of nature in conformity with its laws and thus finally to achieve 'the triumph of art over nature' (*victoria cursus artis super naturam*)."[192] The consequences for human beings' self-understanding and capacity for relating to the Divine are tragic:

> [I]f science as a whole is this and this alone, then it is man himself who ends up being reduced, for the

---

[190] Benedict XVI, *Cor Unum* (Jan. 19, 2013).

[191] Benedict XVI, Address on the Occasion of Christmas Greetings to the Roman Curia (Dec. 20, 2010), http://www.vatican.va/holy_father/benedict_xvi/speeches/2010/december/documents/hf_ben-xvi_spe_20101220_curia-auguri_en.html.

[192] Benedict XVI, *Spe Salvi* (Nov. 30, 2007), 16.

specifically human questions about our origin and destiny, the questions raised by religion and ethics, then
have no place within the purview of collective reason as
defined by "science", so understood, and must thus be
relegated to the realm of the subjective. The subject then
decides, on the basis of his experiences, what he consid-
ers tenable in matters of religion, and the subjective
"conscience" becomes the sole arbiter of what is ethical.
In this way, though, ethics and religion lose their power
to create a community and become a completely personal matter.[193]

The problems with modern science extend to modern technology.
For Benedict, the problem is not with technology in itself, yet there is a
tendency to allow technology's instrumental orientation to guide human action and to succumb to a prideful temptation in the search for
power over nature. "Produced through human creativity as a tool of
personal freedom, technology can be understood as a manifestation of
absolute freedom, the freedom that seeks to prescind from the limits
inherent in things."[194] True freedom comes from recognizing the dependence of humanity on the natural law and God's grace, but technological man is propelled forward in celebration of a false and narrow
freedom of choice and control.

A person's development is compromised, if he
claims to be solely responsible for producing what he
becomes. By analogy, the development of peoples goes
awry if humanity thinks it can re-create itself through

---

[193] Benedict XVI, "Faith, Reason and the University."
[194] Benedict XVI, *Caritas in veritate*, 70.

the "wonders" of technology .... In the face of such Promethean presumption, we must fortify our love for a
freedom that is not merely arbitrary, but is rendered
truly human by acknowledgment of the good that underlies it.[195]

Like modern science, humanity's technology and application of it is
tied closely to an anthropological vision of human nature and its principal end, for "technology is never merely technology. It reveals man
and his aspirations towards development, it expresses the inner tension
that impels him gradually to overcome material limitations."[196] That
"inner tension" may reflect a rational striving toward truth, toward fulfillment of human beings' destiny in loving relationship with God and
stewardship of God's creation, or it may reveal persons' fearful and acquisitive devolution toward sinful concupiscence. "[T]echnological development can give rise to the idea that technology is self-sufficient
when too much attention is given to the 'how' questions, and not
enough to the many 'why' questions underlying human activity. For this
reason technology can appear ambivalent."[197]

Instrumental rationality is focused only on the utility of means for
successful action. Self-giving love grounded in truth, however, is more
desirable than the fragmented efforts at acquisition, possession, and
control. "The dangers of standardization and control, of intellectual and
moral relativism, [are] already clearly recognizable in the erosion of the
critical spirit, the subordination of truth to the play of opinions, the

---

[195] Benedict XVI, *Caritas in veritate*, 68.
[196] Benedict XVI, *Caritas in veritate*, 69.
[197] Benedict XVI, *Caritas in veritate*, 70.

multiple forms of degradation and humiliation of the person's intimacy."[198]

Charles Taylor explains that the Cartesian understanding of experience of the external world, while radically reflexive in the capacity of a person to not only experience their thoughts but examine them objectively, does a kind of violence to the role of personal intentions and understanding in the experience of the object. The "decisive certainty" derided by Benedict, that which propels modern science's technological mission of control over nature, is achieved through a sense of personal objectivity that strives to eliminate any subjective intentions, perspectives, values, and desires from experience of the cosmos:

> Radical reflexivity is central to this stance, because we have to focus on first-person experience in order so to transpose it. The point of the whole operation is to gain a kind of control. Instead of being swept along to error by the ordinary bent of our experience, we stand back from it, withdraw from it, reconstrue it objectively, and then learn to draw defensible conclusions from it. To wrest control from "our appetites and our perceptors", we have to practise a kind of radical reflexivity. We fix experience in order to deprive it of its power, a source of bewitchment and error.[199]

---

[198] Benedict XVI, Address to Participants in a Congress on "Digital Witnesses, Faces and Languages in the Cross-Media Age" (April 24, 2010), http://www.vatican.va/holy_father/benedict_xvi/speeches/2010/april/documents/hf_ben-xvi_spe_20100424_testimoni-digitali_en.html.

[199] Charles Taylor, *The Sources of the Self: The Making of the Modern Identity* (Cambridge, Massachusetts: Harvard University Press, 1989), 163.

Hans Urs von Balthasar describes this as a "hegemony of instrumental rationality" that "reduces nature to the level of brute fact; in principle, therefore, it can dispense with the aspect of personal indebtedness and of goodness; it sees itself simply as an instrument of power and operates as such."[200] Benedict, however, teaches that "the love of God who is Logos" imbues the cosmos with a very personal givenness and goodness, a goodness that is rational, a truth that is both universal and unified and which underlies the orientation of all being to the Creator as final cause and principal end. Humanity's rational experience of that cosmos can only be in the light of a loving relationship with God.

We find a related argument in Benedict's writings about love itself. In the form of love called eros, the person oriented to subjective satisfaction and acquisition of the loved object. This is a kind of instrumental rationality that treats the beloved as a radically separate object. There is another form of love, however, called agape. With agape, the person desires self-expansion through assuming the interests of – perhaps even spiritual communion with – the beloved; the distance between subject and object is bridged.

> Even if eros is at first mainly covetous and ascending, a fascination for the great promise of happiness, in drawing near to the other, it is less and less concerned with itself, increasingly seeks the happiness of the other, is concerned more and more with the beloved, bestows itself and wants to "be there for" the other. The element of agape thus enters into this love, for otherwise eros is impoverished and even loses its own nature. On the other hand, man cannot live by oblative, descending

---

[200] Hans Urs von Balthasar, *Theo-Drama: Theological Dramatic Theory: The Action*, trans. Graham Harrison, vol. 4 (San Francisco: Ignatius Press, 1994), 156.

love alone. He cannot always give, he must also receive. Anyone who wishes to give love must also receive love as a gift. Certainly, as the Lord tells us, one can become a source from which rivers of living water flow (cf. Jn 7:37-38). Yet to become such a source, one must constantly drink anew from the original source, which is Jesus Christ, from whose pierced heart flows the love of God (cf. Jn 19:34).[201]

True love, of course, is love in the grace of God where the person mysteriously gains real freedom by accepting their dependence on God as the "original source." This communion with God, through Christ, is not merely a private relationship of the person and God: "Union with Christ is also union with all those to whom he gives himself."[202] Moreover, the kind of rationality that leads us to Christ is loving and self-expanding, not the selfishly acquisitive and controlling disposition of instrumental rationality.

It is and remains a fundamental word of faith when John—taking up and deepening the creation narrative of the Old Testament—begins his Gospel with the words: "In the beginning was the Logos", the creative reason, the power of the divine knowledge that imparts meaning. It is only from this beginning that one can

---

[201] Benedict XVI, *Deus caritas est*, 7.
[202] Benedict XVI, *Deus caritas est*, 14.

correctly understand the mystery of Christ, in which reason can then be seen to be the same as love.[203]

Benedict also teaches that the problem of instrumental rationality arises in our attempts to control and alleviate suffering.

> We can try to limit suffering, to fight against it, but we cannot eliminate it. It is when we attempt to avoid suffering by withdrawing from anything that might involve hurt, when we try to spare ourselves the effort and pain of pursuing truth, love, and goodness, that we drift into a life of emptiness, in which there may be almost no pain, but the dark sensation of meaninglessness and abandonment is all the greater.[204]

Moreover, this struggle in regard to one's individual life is directly relevant to how we attend to others' suffering. The person who lives in union with Christ knows that suffering is a profoundly meaningful aspect of life, and that love for one's neighbor includes helping them to embrace and pray over their suffering. It is not a utilitarian emphasis on calculation of the best means to control and remove the unwanted material conditions.

> Those who work for the Church's charitable organizations must be distinguished by the fact that they do not merely meet the needs of the moment, but they

---

[203] Joseph Ratzinger, *A Turning Point for Europe? The Church in the Modern World*, trans. Brian McNeil, second ed. (San Francisco: Ignatius Press, 1994), 110-111.

[204] Benedict XVI, *Spe Salvi*, 37.

dedicate themselves to others with heartfelt concern, enabling them to experience the richness of their humanity.[205]

It is the virtuous Christian, skilled in the art of suffering as well as in experiencing joy, who can most help others, for:

> the individual cannot accept another's suffering unless he personally is able to find meaning in suffering, a path of purification and growth in maturity, a journey of hope. Indeed, to accept the 'other' who suffers, means that I take up his suffering in such a way that it becomes mine also.[206]

Pope Francis:

Pope Francis balances a positive view of the opportunities which modern technology present for integral human development with a grave concern about the instrumental rationality that is inherent in its use. He states that "technology itself 'expresses the inner tension that impels man gradually to overcome material limitations.' Technology has remedied countless evils which used to harm and limit human beings."[207] At the same time, the development and use of advanced technology is too frequently the source of domination of other persons:

> We have to accept that technological products are not neutral, for they create a framework which ends up conditioning lifestyles and shaping social possibilities

---

[205] Benedict XVI, *Deus caritas est*, 31.
[206] Benedict XVI, *Spe salvi*, 38.
[207] Francis, *Laudato si'* (May 24, 2015), 102.

along the lines dictated by the interests of certain powerful groups. Decisions which may seem purely instrumental are in reality decisions about the kind of society
we want to build.[208]

Going beyond concerns about social and material welfare, Francis
demonstrates that the societal focus on development of technologies is
a spiritual and moral problem. He calls this the "technocratic paradigm," by which the instrumental rationality of utility, efficiency, and
control becomes a dominant way of behavior in human relations. This
paradigm is primarily about the exercise of power.

This paradigm exalts the concept of a subject who,
using logical and rational procedures, progressively approaches and gains control over an external object. This
subject makes every effort to establish the scientific and
experimental method, which in itself is already a technique of possession, mastery and transformation. It is
as if the subject were to find itself in the presence of
something formless, completely open to manipulation.
… Human beings and material objects no longer extend
a friendly hand to one another; the relationship has become confrontational.[209]

As with the alienation discussed by Pope St. John Paul II and the
false subject-object dichotomy highlighted by Pope Benedict, this technocratic paradigm is a cultural phenomenon that requires a vigorous
challenge to eradicate it; "the idea of promoting a different cultural

---

[208] Francis, *Laudato si'*, 107.
[209] Francis, *Laudato si'*, 106.

paradigm and employing technology as a mere instrument is nowadays inconceivable."[210] The technocratic paradigm is not inevitable, but it is a persistent and intensifying phenomenon.

> Technology tends to absorb everything into its ironclad logic, and those who are surrounded with technology "know full well that it moves forward in the final analysis neither for profit nor for the well-being of the human race", that "in the most radical sense of the term power is its motive – a lordship over all". As a result, "man seizes hold of the naked elements of both nature and human nature". Our capacity to make decisions, a more genuine freedom and the space for each one's alternative creativity are diminished.[211]

Aside from the characterization of the technocratic paradigm as a strongly powerful force in society, Francis announces a hopeful message of the capacity of persons to overcome such a temptation. "A better world is possible thanks to technological progress if this is accompanied by an ethic inspired by a vision of the common good, an ethic of freedom, responsibility, and fraternity, capable of fostering the full development of people in relation to others and to the whole of creation."[212] This capacity is most often expressed in terms of social welfare and community, the common good, and Francis is strikingly different from his predecessors in his lesser emphasis on contemplation of being, truth, and philosophical renewal. When Francis laments the loss of a vision of the

---

[210] Francis, *Laudato si'*, 108.

[211] Francis, *Laudato si'*, 108.

[212] Francis, Address to the Participants in the Seminar "The Common Good in the Digital Age" (Sep. 27, 2019).

totality of being through the "fragmentation" and specialization of technical knowledge, it is because "This very fact makes it hard to find adequate ways of solving the more complex problems of today's world, particularly those regarding the environment and the poor; these problems cannot be dealt with from a single perspective or from a single set of interests."[213] He also indicates that "What is needed is to situate scientific and technological knowledge within a broader horizon of meaning, and thus to avert the hegemony of a technocratic paradigm."[214] Here, the focus is on a "horizon of meaning," reflecting both the phenomenological language and emphasis on a collective but subjectively experienced "meaning" rather than "truth" as an alternative to instrumental, interest-based domination. This is not to say that Francis disregards truth as crucial to the Christian life but rather displays a different overall emphasis in his discussion of instrumentally-oriented technology. As the saturation of the culture by modern technologies accelerates, so do warnings of the dire material and spiritual consequences if scientific and technical progress are not placed in the proper context of a God-centered morality.

Francis ties his ecological concerns to problems with the way our culture thinks of and treats human persons. Responsible ecological stewardship includes a radical change in the anthropology by which humanity views itself.

> This situation has led to a constant schizophrenia, wherein a technocracy which sees no intrinsic value in lesser beings coexists with the other extreme, which sees no special value in human beings. But one cannot

---

[213] Francis, *Laudato si'*, 110.

[214] Francis, "Human. Meanings and Challenges," address to the General Assembly of the Pontifical Academy for Life (Feb. 12, 2024).

prescind from humanity. There can be no renewal of our relationship with nature without a renewal of humanity itself. There can be no ecology without an adequate anthropology.[215]

A willingness to alter our understanding of humanity to suit the efficiency, control, and task-oriented demands of technocratic rationality leads to relativism in both our anthropology and our moral values.

> [T]he practical relativism typical of our age is "even more dangerous than doctrinal relativism". When human beings place themselves at the centre, they give absolute priority to immediate convenience and all else becomes relative. Hence we should not be surprised to find, in conjunction with the omnipresent technocratic paradigm and the cult of unlimited human power, the rise of a relativism which sees everything as irrelevant unless it serves one's own immediate interests. There is a logic in all this whereby different attitudes can feed on one another, leading to environmental degradation and social decay.[216]

Biblical Warnings

The Call of the Cross:

Jesus does not hide the paradoxical nature of his call to his disciples to follow him in his passion and sacrifice: "And he said to all, 'If anyone would come after me, let him deny himself and take up his cross daily

---

[215] Francis, *Laudato si'*, 118.
[216] Francis, *Laudato si'*, 122.

and follow me. For whoever would save his life will lose it, but whoever loses his life for my sake will save it."' (Luke 9:23-24) There is an opposition drawn between biological life and the life of love found in devoted relationship with Christ. The apparent good of biological life, however, is itself derived – in the perspective of the human person – from the utility of life which is the capacity and opportunity to experience the pleasures and challenges of lived, active being. To accept suffering and sacrifice, indeed to embrace it, is an apparently irrational violation of the instrumentally rational pursuit of utility, efficiency, effectiveness, and maximization of finite goods. Jesus challenges us with a relational and holistic rationality that is incompatible with instrumental rationality.

This call to resist the temptation of instrumentality is further displayed in Jesus' challenge to give up our dependence on wealth: "If you would be perfect, go, sell what you possess and give to the poor, and you will have treasure in heaven; and come, follow me" (Matthew 19:21). Hoarding riches will not result in happiness, security, or eternal life.

> And he told them a parable, saying, "The land of a rich man produced plentifully, and he thought to himself, 'What shall I do, for I have nowhere to store my crops?' And he said, 'I will do this: I will tear down my barns and build larger ones, and there I will store all my grain and my goods. And I will say to my soul, "Soul, you have ample goods laid up for many years; relax, eat, drink, be merry."' But God said to him, 'Fool! This night your soul is required of you, and the things you have prepared, whose will they be?'" (Luke 12:16-21)

There is a fundamental difference in the being of a person in relation to money or to God. A person can have a relationship with money, and by extension a relationship with a plethora of material and psychological goods, but this relationship is one of opposed subject and object – radically contingent, insecure, and one-sided. A person cannot truly be their wealth, they can only own it, use it, wield it, count it, and treasure it for its instrumental effectiveness as a means to limited goods. On the other hand, the relationship with God, through Christ, is one of loving union and self-giving, in which human persons and God live out their essential natures – "life" here taking on a higher meaning. The Christian leap from a commitment to wealth to a reverential commitment to God is not merely an exchange of beloved objects, but a commitment to the very being of humanity itself. In the parable, the rich man tragically loses his soul; he no longer has being, no longer *is*.

Idolatry:

A concern over idolatry permeates the Old Testament, especially in the Pentateuch. While the central issue in idolatry is the simple displacement of reverence and love for God to another being or image, the motivation for engaging in idolatry is quite often instrumental; the object of worship gains its status in the mind of the worshipper due to its function as means of attaining or acquiring desired goods. Perhaps the best example of this kind of idolatry is found in Jesus' warning: "You cannot serve God and mammon" (highlighted also in the Catholic Catechism passage on idolatry).[217] Mammon – money or wealth – is a means of securing many goods, including material pleasures and power, but on a deeper level it provides psychological goods such as security, prestige, and confirmation of success in practical efforts. It is the

---

[217] Matthew 6:24; Catholic Church, *Catechism of the Catholic Church*, 2nd ed. (Vatican: Libreria Editice Vaticana, 1997), 2113.

fungibility of money that makes it so easily worshiped. Like money, one's own work and its products can be idolized, as described vividly in Isaiah 44:12–17.

Pope Francis, in one of his weekly general audiences, taught that we often make idols out of the objects that are means to our subjective and limited satisfaction.

> The idol in reality is a projection of self onto objects or projects. Advertising, for example, uses this dynamic: I cannot see the object itself but I can perceive that car, that smartphone, that role — or other things — as a means of fulfilling myself and responding to my basic needs. And I seek it out, I speak of it, I think of it: the idea of owning that object or fulfilling that project, reaching that position, seems a marvelous path to happiness, a tower with which to reach the heavens (cf. Gen 11:1-19), and then everything serves that goal.[218]

Idolatry is therefore intimately connected to our anxiety about successfully utilizing available means to secure that which we feel a dependence on. "Everything stems from the inability to trust above all in God, to place our safety in Him, to let Him give true depth to the desires of our heart. Without God's primacy one easily falls into idolatry and is content with meager assurances."[219] As for Evagrius and the Desert Fathers, temptation to reduce one's attitude and activities to desperate pursuit of finite goods is heightened by anxiety over the loss of such goods. We all live at times in the desert. "What is the desert? It is a place where

---

[218] Francis, General audience (August 1, 2018), https://www.vatican.va/content/francesco/en/audiences/2018/documents/papa-francesco_20180801_udienza-generale.html.

[219] Francis, General audience (August 1, 2018).

uncertainty and insecurity reign – there is nothing in the desert – where there is no water, there is no food and there is no shelter."[220]

Idols also may not directly displace God as the center of our attention but may appear to offer a false promise of control over the Divine power. In the Old Testament, God tells Moses to "make a fiery serpent, and set it on a pole; and every one who is bitten, when he sees it shall live. So Moses made a bronze serpent, and set it on a pole; and if a serpent bit any man, he would look at the bronze serpent and live" (Num. 21:8–9). In this scenario, then, the bronze serpent was a finite means for utility, but the attitudes and actions of the believers were expected to be couched in deep trust and orientation to love of the Lord. The situation changed, however, when the bronze serpent became a utilitarian end in itself, leading to worship of the serpent as a god. King Hezekiah recognized the conversion of appropriate means-end logic into an instrumental rationality that literally worshipped the means and obscured the principal end; the serpent was therefore destroyed (2 Kgs. 18:4). In our society that is culturally and economically dependent on complex systems of financial instruments, wage labor, and modern technologies, the option of simply destroying our idols is much less feasible or palatable.

Babel:

The account of Babel (Gen. 11:4) – its people's audacious construction of "a city and a tower" (perhaps a pyramid-like ziggurat) that reached toward the heavens and their subsequent linguistic and geographic dispersion – is an important bridge between the creation accounts in Genesis and the New Testament experience of Christ's apostles on Pentecost, yet it is also a stark warning about instrumental

---

[220] Francis, General audience (August 1, 2018).

rationality, especially in its entrenched, ideological form. Pope Francis spoke of this connection at a general audience in 2023, later repeated similarly in a communication to the General Assembly of the Pontifical Academy for Life:

> The account of the city of Babel and its tower comes to mind (cf. Gen 11:1-9). It narrates a social project that involves sacrificing all individuality to the efficiency of the collective. Humanity speaks only one language — we might say that it has a "single way of thinking" — as if enveloped in a kind of general spell that absorbs the uniqueness of each into a bubble of uniformity. Then God confuses the languages, that is, He re-establishes differences, recreates the conditions for uniqueness to develop, revives the multiple where ideology would like to impose the single.[221]

Here, Francis speaks of not only the narrowness of an ideological "way of thinking," but characterizes such thinking as a devaluation of individual human nature for the sake of mass effort and efficiency; the people of Babel had become mere means in their effort to elevate themselves. Francis continues:

> This story really does seem topical: even today, cohesion, instead of fraternity and peace, is often based on ambition, nationalism, homologation and techno-economic structures that inculcate the conviction that God is insignificant and useless: not so much because one

---

[221] Francis, General audience (Nov. 29, 2023), https://www.vatican.va/content/francesco/en/audiences/2023/documents/20231129-udienza-generale.html.

seeks more knowledge, but above all for the sake of more power. It is a temptation that pervades the great challenges of today's culture.[222]

According to Francis, then, the contemporary society is like Babel in its orientation not only to challenge God's supremacy but to do so primarily in reference to a distorted perception of the utility of God's reign. Our omnipotent God, however, refuses to allow such a single mindset, causing dispersion and inefficient variability among the people. You might say it is the assertion of natural law through expression of human freedom, which is most fully expressed in pursuing one's principal, teleological end. "In this way, human beings would come face to face with their limitations and vulnerability, and be challenged to respect differences and to show concern for one another."[223]

It was the "builders" in the Genesis account who had one language and were subsequently dispersed, and these builders are described as technical masters of the new process of brick-making; the unity of language was intimately bound up with the technological and instrumental effectiveness of the builders, and this pride-directed pursuit of utility was destroyed with the linguistic scattering.[224] The building of the ziggurat in Babel was essential to the builders' hope to manipulate a pagan deity with worship and supplications. "The tyrannical twosome of self-sufficient technology and self-serving religion represents a human bid for self-achieved security. Herein lie the origins of Empire-building."[225]

---

[222] Francis, General audience (Nov. 29, 2023).

[223] Francis, "Human. Meanings and Challenges."

[224] Paul S. Evans, "Creation, Progress, and Calling: Genesis 1-11 as Social Commentary," *McMaster Journal of Theology and Ministry* (2011), 90-92.

[225] Evans, "Creation, Progress, and Calling," 93.

In The City of God, Augustine perceived the construction of the tower of Babel as a prideful effort at domination.[226] God's scattering of peoples and languages is therefore a rebuke of this domination. The punishment of the builders described in the Bible also directly ties the offending sin to the design and construction of the tower. "But what was the nature of the punishment? As the tongue is the instrument of domination, in it pride was punished; so that man, who would not understand God when He issued His commands, should be misunderstood when he himself gave orders."[227] Augustine's warning against technical domination and its sinful failure to trust in God's providence reflects a passage in Isaiah 50:10-11.

> Who among you fears the Lord and obeys the voice of his servant? Let him who walks in darkness and has no light trust in the name of the Lord and rely on his God. Behold, all you who kindle a fire, who equip yourselves with burning torches! Walk by the light of your fire, and by the torches that you have kindled! This you have from my hand: you shall lie down in torment.

Proverbs:

The biblical Proverbs contain fundamental and consistent guidance in regard to instrumental rationality, without naming the disposition itself. Believers are encouraged to pursue intermediate and finite goods like wealth and prosperity, but the principal end of humanity in a reverent and obedient relationship with God requires that the virtuous manner in which means are applied in attaining such finite goods matters greatly. Utility is not the only criteria for the use of means and a

---

[226] Augustine, *The City of God* XVI, 4.
[227] Augustine, *The City of God* XVI, 4

person's relationship with others. Even more profound is the message that virtue is not an exogenous or imposed criteria for persons' evaluation of means, but is a crucial element of success in attaining even the intermediate goods for their happiness.

Proverbs 8 includes an extended, anthropomorphized description of wisdom. Verse 12 states: "I, Wisdom, dwell with prudence, and useful knowledge I have." Wisdom, which is generally the kind of knowledge possessed by the virtuous and divinely blessed person, is not divorced from prudence, which is right reason in practical matters. Such right reason pertains to the right pairing of means and ends for utility. In the modern world, the treatment of moral and transcendental matters as irrelevant to practical calculation of means is quite different from the picture of wisdom offered in Proverbs 8:12. In the proverb, not only does wisdom offer clear benefits, but it is enabled by maintaining a focus on the true or principal end when choosing means toward intermediate purposes. Truth – not merely factual or empirical, but transcendent, holistic, and moral – is therefore crucial for right discernment of the means for practical action. This truth engages the criteria of authority (specifically, Divine authority found in revelation) in appraising the evidence related to one's calculation of appropriate means.

We see this integrated concept of wisdom and prudence in the proverbs that equate prudence, discipline, and skill with the way of righteousness and prosperity. For example:

> Wealth won quickly dwindles away, but gathered little by little, it grows. (Proverbs 13:11)

> The wisdom of the shrewd enlightens their way, but the folly of fools is deceit. (Proverbs 14:8)

Do you see those skilled at their work? They will stand in the presence of kings, but not in the presence of the obscure. (Proverbs 22:29)

The slack hand impoverishes, but the busy hand brings riches. (Proverbs 10:4)

Sloth does not catch its prey, but the wealth of the diligent is splendid. (Proverbs 12:27)

On the other hand, how wealth is acquired – that is, what character the subject displays when in pursuit of wealth – matters just as much as the strict utility of the means. "Ill-gotten treasures profit nothing, but justice saves from death" (Proverbs 10:2).

An unjust or non-virtuous pursuit of wealth is not only immoral, but will ultimately fail to achieve the pursuit of the good desired. Actively unjust or deceitful attempts at gaining wealth will lead to ruin:

Misers hurry toward wealth, not knowing that want is coming toward them. (Proverbs 28:22)

Whoever amasses wealth by interest and overcharge gathers it for the one who is kind to the poor. (Proverbs 28:8)

Balance and scales belong to the LORD; every weight in the sack is his concern. (Proverbs 16:11)

Trying to get rich by lying is chasing a bubble over deadly snares. (Proverbs 21:6)

The Proverbs also indicate that a person's own cleverness and especially planning are not enough to ensure happy returns on their efforts. "It is the LORD's blessing that brings wealth, and no effort can substitute for it" (Proverbs 10:22). There is no purely epistemic or scientific knowledge of the world that will enable a person to securely achieve lofty goals without the deep wisdom attained through an intimate relationship with God and His revelation. "The human heart plans the way, but the LORD directs the steps" (Proverbs 16:9). In the terms associated with instrumental rationality, calculation of means for an intermediate end is simply not secure without orientation of both action and ends toward fulfillment of human persons' principal end.

> Plans are made in human hearts,
> but from the LORD comes the tongue's response.
> All one's ways are pure in one's own eyes,
> but the measurer of motives is the LORD.
> Entrust your works to the LORD,
> and your plans will succeed. (Proverbs 16:1-3)

It at first seems ironic that the Proverbs also teach us that planning and diligence in following plans are virtuous actions. This is clearly outlined in Proverbs 3:21-26:

> My son, do not let these slip from your sight:
> hold to deliberation and planning;
> So will they be life to your soul,
> and an adornment for your neck.
> Then you may go your way securely;
> your foot will never stumble;

> When you lie down, you will not be afraid,
> when you rest, your sleep will be sweet.
> Do not be afraid of sudden terror,
> of the ruin of the wicked when it comes;
> For the LORD will be your confidence,
> and will keep your foot from the snare. (Proverbs
> 3:21-26)

The goodness of planning is also reaffirmed in Proverbs 12:5 and 24:27. Planning is prudent, and it entails both effort and intention toward excelling at one's own work. Failing to plan, on the other hand, is not only lazy, but it increases the likelihood that a person living in an agrarian society will succumb to misfortune and then extract unfair benefits from the generosity of others.

The Proverbs, in fact, express their greatest disdain for the lazy, maliciously clever, and imprudent sluggard. "To the LORD the devious are an abomination, but the upright are close to him" (Proverbs 3:32) Even an ant is more wise than a sluggard:

> Go to the ant, O sluggard,
> study her ways and learn wisdom;
> For though she has no chief,
> no commander or ruler,
> She procures her food in the summer,
> stores up her provisions in the harvest. (Proverbs
> 6:6-8)

One passage indicates that a primary motivation of the sluggard's paralysis is fear of potential and imagined circumstances. "The sluggard says, 'A lion is outside; I might be slain in the street'" (Proverbs 22:13). The sluggard suffers from a lack of prudent planning and from unwise mistrust of both his dignity as a worker and the Lord's providence. These cause the sluggard to hide away from extending his efforts. Here, we have a connection between the disposition of instrumental rationality, the anxiety of acedia, and the sinful depression of activity found in sloth. Lack of wisdom and an inordinate focus on subjective, factual data generates a failure in prudent calculation of the available means for success, leading to paralysis and even rejection of the intermediate end.

Chapter 4

Instrumental Rationality and AI

Earlier in this work, Max Weber's concept of rationalization was briefly introduced. This is a process in which a society or social unit adopts regular, enduring patterns of action that favor one or more types of rationality. Social patterns and structures, including material incentives and sanctions, may reinforce certain motivations and thought patterns so that the instituted type of rationality progressively dominates additional actions and spheres of social life. As explained earlier, there is nothing necessary, inevitable, or automatic about any type or process of rationalization, nor is it historically or culturally universal; rationalization is a descriptive term that denotes the ossified and often institutionalized persistence of a particular rational disposition in a social group.

While Weber was focused largely on cognitive theories, concepts, and ideologies in his descriptions of rationalization, he did give at least one important example of the role of material incentives and constraints in the case of bureaucratic organizations that institute what he calls a formal rationality (we might think of pay scales, office arrangements, structured resource distribution, and military disciplinary actions). Weber's notion of formal rationality refers, in a sociological sense, to the systemic or persistent application of rules to regulate action. It is a description of a pattern of behaviors, or the disposition to behave as such; it is not a kind of reasoning as a purely cognitive power. This notion of "formal," applied by Weber to bureaucratic structures and here to physical machines, is not intended to refer to the Thomistic philosophical understanding of immaterial form. Perhaps a better term for formal rationality would be institutional rationality or simply rules-based rationality.

In this work, emphasis will be placed on the material incentives and constraints that characterize AI technologies and lead to instrumental rationalization, as well as the more ideological or theoretical influences of those technologies. In fact, the very structure of AI as a material system of machines and as a conceptual operating system will be particularly important to the instrumental rationalization of society. More specifically, the design of AI and the active environment of its users enable, constrain, incentivize, or penalize particular ways of life that, in turn, are characterized by a heavy emphasis on instrumental rationality.

A classic example of material incentives and constraints generated by the design of a technology comes from a story told by Friedrich Kittler about the philosopher Friedrich Nietzsche's encounter with an early form of typewriter.[228] Nietzsche had written many long works in prose or a more poetic form, but in his later years succumbed to increasing loss of his eyesight. To help him write, he purchased a Hansen Writing Ball typewriter. As Kittler tells it, the typewriter was structured inefficiently, perhaps because emphasis was placed on easily carrying it from place to place (which we can assume meant it was smaller and simpler). Specifically, Nietzsche had difficulty seeing what it was that he had written because the pages were hidden as they passed through the typewriter, and mistakes might lead to laborious re-writing of passages. To make a long story short, Nietzsche adjusted his writing to focus on the use of pithy aphorisms that expressed deep truths or principles in just a sentence or two. Not only his writing, but his thought process turned to acerbic, clever statements that were often disconnected from longer

---

[228] Friedrich A. Kittler, *Gramophone, Film, Typewriter*, translated by Geoffrey Winthrop-Young and Michael Wutz (Stanford: Stanford University Press, 1999), 200-208.

arguments. Concerns over efficiency and brevity therefore converted Nietzsche's rational process away from more discursive, deductive arguments to declarative insights.

AI technology similarly favors instrumental rationality in a variety of ways. Consider the the very term "artificial intelligence." One survey of more than seventy prominent sources that describe AI found that they defined intelligence as a characteristic of agents that are intending to achieve measurable goals in an external environment through efficient and effective data collection, learning, and adaptation.[229] This is a very restrictive understanding of intelligence that does not include contemplative abilities, moral intentionality, virtuous growth of the agent, and apprehension of principles through synderesis and their implications for practical action. Although there may be some development (really imitation) of human reasoning in AI models, this understanding of intelligence is hardly conducive to theoretical insight and deduction, analogical and abductive reasoning (i.e., common sense), or apprehension and facilitation of goodness, however defined.[230]

Brian Cantwell Smith, for example, tackles these instrumental notions of intelligence:

> [I]f the topic is intelligence, and its importance to theology, to things that matter, to our place in the universe, then the analysis should consider systems as wholes, undistracted by low-level details about ingredients and inner constitution.

---

[229] Shane Legg and Marcus Hutter, "A Collection of Definitions of Intelligence," in *Frontiers in Artificial intelligence and Applications* 157 (2007), 17-24.

[230] Taylor Webb, Keith J. Holyoak, and Hongjing Lu, "Emergent Analogical Reasoning in Large Language Models," *Nature Human Behavior* 7 (2023), 1526-1541, https://doi.org/10.1038/s41562-023-01659-w

In recent work, I introduce one such distinction, designed to satisfy both criteria: between what I call (a) reckoning, a kind of calculative intelligence exemplified by systems not themselves responsible for the relation between their symbols and data structures and information and the situations or worlds that those structures are about, and (b) judgment, a form of deliberative intelligence, grounded in existential commitment and responsible action, that is both appropriate and ethically responsible to the situation in which it is deployed. By "judgment," that is, I mean the sort of thing we get at when we say that someone "has good judgment."

In terms of this distinction, my overall claims are two. First, "intelligence," on its own, is too broad and unruly a concept to be useful in asking ultimate questions; insight requires a finer-grained understanding of its kinds. Second, our informal notion of intelligence is broader and deeper than current technical discussions suggest; full-blooded intelligence, in my view, requires genuine judgment.[231]

Christos Kyriacou draws on Aristotelian virtues theory to identify four characteristics of human intelligence that are inevitably missing from what he calls the "mechanical instrumental rationality" of AI.[232] A human intelligence engages in judgments based on both absolute moral

---

[231] Brian Cantwell Smith, *The Promise of Artificial Intelligence: Reckoning and Judgment* (Cambridge, MA: MIT Press, 2019).

[232] Christos Kyriacou, "Artificial Moral Intelligence and Computability: An Aristotelian Perspective," *AI and Ethics* (2024), https://doi.org/10.1007/s43681-024-00543-1

principles and the use of categorical reasoning, autonomously directs acts according to deduced reasons, and draws upon affective experience. The mechanical intelligence, however, must act on contingent data and conditions with either programmed or means-end calculation of contingent ends. At least in the normative sphere, it seems that mechanical intelligence is missing significant aspects necessary for fully autonomous, successful normative reasoning.

Intelligence is a spiritual power of the human person in the image of God. The Dicastery for the Doctrine of the Faith expresses this eloquently in the 2025 document *Antiqua et Nova*:

> A proper understanding of human intelligence, therefore, cannot be reduced to the mere acquisition of facts or the ability to perform specific tasks. Instead, it involves the person's openness to the ultimate questions of life and reflects an orientation toward the True and the Good. As an expression of the divine image within the person, human intelligence has the ability to access the totality of being, contemplating existence in its fullness, which goes beyond what is measurable, and grasping the meaning of what has been understood. For believers, this capacity includes, in a particular way, the ability to grow in the knowledge of the mysteries of God by using reason to engage ever more profoundly with revealed truths (*intellectus fidei*). True intelligence is shaped by divine love, which "is poured forth in our hearts by the Holy Spirit" (Rom. 5:5). From this, it follows that human intelligence possesses an essential *contemplative* dimension, an unselfish openness to the

True, the Good, and the Beautiful, beyond any utilitarian purpose.[233]

AI technology is oriented around a decidedly artificial intelligence;
it is borne and manifested by machines. These machines are always
purely material means for accomplishing the goals of human persons.
The structure of these machines is a series of physical means and software programs that rely for their meaningfulness on their designated,
instrumental use and the mental models that guide their manufacturing
and development. Machines are tools. They represent a particular organization of separable, potentially independent parts. Despite ever
greater complexity, continuity of operations, and autonomy, the machines are intended to carry out tasks, general purposes, or imitations
of biological life in a way that suits their designers. The capacity and
structure of a machine as well as the associated physical properties of
its components also define, limit, incentivize, change, and generate purposes that are willed by the human participants. Those parameters can
be more or less encouraging of an instrumental nature of ends.

One uniquely prominent characteristic of computer machines is
that human meanings are attributed to the inputs (data), operating
schemas, and output of the machines; users and even the designers increasingly substitute these applied meanings for material descriptions
of the inputs, schemas, and outputs in a way that makes sense to themselves and others. For example, the "on" or "off" positions of binary
switches in the central processing unit (CPU) are understood as indicating "yes" and "no" in logical sequences, arrangements of the material
structure to generate particular outputs in response to particular inputs

---

[233] Dicastery for the Doctrine of the Faith and Dicastery for Culture and Education, *Antiqua et Nova* "On the Relationship between Artificial Intelligence and
Human Intelligence" (January 28, 2025), 29.

become known as "programs," and the digital outputs of a word processing program become "words" and "sentences" along with their further stratified levels of meaning and communication. Even the currently
overused word "algorithm" betrays a false sense of unity and thought
that somehow transcends the jumble of computer components and hu-
man-labeled data that underlie any AI system's operations. Ian Bogost
declares:

> The algorithm has taken on a particularly mythical
> role in our technology-obsessed era, one that has al-
> lowed it to wear the garb of divinity. Concepts like "al-
> gorithm" have become sloppy shorthands, slang terms
> for the act of mistaking multipart complex systems for
> simple, singular ones. Of treating computation theolog-
> ically rather than scientifically or culturally.[234]

AI represents a change in the very same computer machines where
human concepts like learning, thinking, and choosing are applied to the
operations of the machines. Most importantly, many perceive the machines or networks of machines to be agents that have at least some capacity for autonomous initiation of processes related to thought, output,
and sometimes action including physical movement that navigates or
affects the external world. Imagination of the possibility of full agency
(i.e. willed and free choice) for AI programs or AI-governed devices is
rather easy to entertain once a limited concept of agency is applied to
computers or the programs that govern them. Even so, contemporary
imaginations of a future artificial general intelligence (AGI) ironically

---

[234] Ian Bogost, "The Cathedral of Computation," *The Atlantic* (January 15,
2015), https://www.theatlantic.com/technology/archive/2015/01/the-cathedral-
of-computation/384300/.

retain the instrumental character of the machines – now focused on the machines' self-interest – rather than anything resembling the diverse personalities and mental or spiritual powers of human beings.

Before proceeding, it will help to more formally define AI and its specific variants or techniques represented by machine learning, neural networks, deep learning, and generative AI. There is no canonical definition of AI, but it generally concerns the science and the resulting programs that give machines the capacity to emulate the powers of human thought. When originally coined, the term was defined in 1955 as "making a machine behave in ways that would be called intelligent if a human were so behaving."[235] Since then, there have been varying understandings of just what AI refers to and what research and goals are appropriate. One popular textbook defines AI as "a system's ability to correctly interpret external data, learn from it, and use that learning to achieve specific goals and tasks through flexible adaptation."[236] Another major textbook by Stuart Russell and Peter Norvig avoids any single definition of AI but instead describes four research programs or "approaches."[237] The Turing Test approach attempts to develop a wide array of actions – both mental and physical – in machines that are indistinguishable from those of human beings, as determined by human observers. The core capacities are natural language processing, knowledge representation, automated reasoning, learning, vision and speech recognition, and controlled robotic motion. Terms such as reasoning and learning are, of

[235] John McCarthy, M. L. Minsky, Nathaniel Rochester, C. E. Shannon, "A Proposal for the Dartmouth Summer Research Project on Artificial Intelligence" (1955), http://www-formal.stanford.edu/jmc/history/dartmouth/dartmouth.html

[236] Michael Haenlein and Andreas Kaplan, "A brief history of artificial intelligence: On the past, present, and future of artificial intelligence," *California Management Review* 61, no. 4 (2019), 5.

[237] Stuart Russell and Peter Norvig, *Artificial Intelligence: A Modern Approach*, 4th edition (Hoboken: Pearson, 2021), 1-4.

course, quite varied in their definition and implications for the scope of a machine's resemblance to human mental powers. The cognitive modeling approach focuses more on mental capacities by attempting to understand human cognition and model it arithmetically or symbolically so that a computer machine might reproduce those capacities. The "laws of thought" approach attempts to develop reasoning (i.e. logical) abilities in machines so that they can perceive, evaluate, understand, and predict. The expectations and objects for each power vary considerably among the researchers and developers. Finally, the rational agent approach expands beyond the goal of inferential powers to making machines act rationally, which is to "do the right thing," given the programmed aims and the circumstances of action.

The applied definition of rationality certainly matters here, and Russell and Norvig provide a definition that attempts to synthesize most AI developers' understanding of a rational agent oriented to producing the right consequences of its actions: "For each possible percept [i.e., perceptual experience] sequence, a rational agent should select an action that is expected to maximize its performance measure, given the evidence provided by the percept sequence and whatever built-in knowledge the agent has."[238] We might note that this is also an expression of means-end reasoning – or better, calculation – that focuses on the selection of intermediate means to achieve a limited goal. The performance measure is an assessable proxy for the developers' purpose, yet it should not be confused with the purpose itself, since changing circumstances or meanings of the machine's actions could inspire the developer or user to alter their original goal, and many goals are not easily operationalized as a set of measurable events and indications. As I have written elsewhere:

---

[238] Russel and Norvig, *Artificial Intelligence*, 40.

It is crucial to note that the process of instrumental rationality is no longer simply carrying out calculations and actions that apparently achieve the meta-goal, but the AI-enabled machine engages in largely hidden layers of calculations, algorithms, and redefinition of data values and their relations in the process of evaluating means and action paths that will most successfully meet the meta-goal. It is means-end calculation at a hyper-level of intellection.[239]

Whenever there is uncertainty about the environment and the effects of action, the AI agent can only maximize expected performance (an average of probabilities and utilities of possible outcomes) modeled internally as a best guess of future states and adjusted iteratively. The "problem" to be solved is a task environment that helps to facilitate the solution – the meta-goal modeled as an external state. Any intermediate goals that potentially conflict are weighed and chosen according to the AI system's algorithmic utility function (its mathematically expressed, internal measure of success).

The knowledge of the machine is an array of measurable or programmed data that is internally related according to a programmed schema (which may be altered by further acquired data), but it is not equivalent to the meaningful, communicable, and contextual pre-understanding of the world that a human being might possess and progressively reform. Even with the most ambitious AI programs, there is little recognition of the human process of imagining, referencing, refining, and discursively negotiating the meanings that are attributed to data, labels, and classifications; the meaning of the data is related solely

---

[239] Reilly, "How Artificial Intelligence Technology Encourages the Vice of Acedia," 163.

to the utility of calculations or actions that achieve the desired output. The data is acquired only through programming, physical sensors, and user actions or inputs. Also, the meanings of data, labels, and their relations are interpreted in a manner that they can be represented by and emerge from mathematical calculations; this excludes much of reality from the data and modeling. The AI goal is also limited by its expression in terms of consequences of the machine's calculations and action; a self-giving, sympathetic, contemplative, purely exploratory (as in aimless musing), reverent, or emotional expression of the goal is, at best, unlikely for such machines. The general orientation appropriate for a developer or user of the AI-governed machine is therefore one that emphasizes the right calculation of intermediate means and actions and the matching of feasible goals rather than significant consideration of the appropriateness, morality, or fulfilling potential of the goals themselves. This is a disposition characterized by instrumental rationality.

One important capacity of an AI-governed machine is the ability to enhance or revise the schema of information and subsequently the actions of the machine, based on experience of new information. Russell and Norvig describe the process of Machine Learning (ML) by which "a computer observes some data, builds a model based on the machine learning data, and uses the model as both a hypothesis about the world and a piece of software that can solve problems."[240] This learning can occur when programmers define particular labels for data and the machine builds upon its developer-controlled experience of labeled information – essentially forming maps or functions of relations between inputs and outputs – to further predict the correct application of labels to new data (supervised learning), or it may occur in the absence of labels through inductive inference and recognition of patterns based on

---

[240] Russel and Norvig, *Artificial Intelligence*, 651.

experience of new data (unsupervised learning). With reinforcement learning, the machine is programmed to recognize and optimize certain "rewards" (or avoid certain "punishments") that guide its pursuit of right actions and increasingly complex memorization of those experiences in new contexts.

Another type or process of AI is deep learning, in which artificial neural networks (ANNs) act as hypotheses with multiple connections between representative nodes (always instantiated as electronic pulses within physical components of the machine) that are further organized as layers of connections hierarchically ordered. The nodes (or "units") of the ANN are variables of either data or actions, and the connections or interactions between nodes – and between layers of connected nodes – can be arithmetically weighted to influence the overall results of calculations. The machine proceeds from the initial inputs through successive layers or iterations of decisions (perhaps even "backward" in more complex, recurrent networks) that are each represented by an array of logically or mathematically connected variables. ANNs are particularly useful for machine learning that is characterized by a very large number of input variables (as with high-dimensional images) and for a longer series of decisions and computations extending between the inputs and outputs. The result is a machine that can act with less dependence on programmed algorithms and more contingently and strategically (with knowledge of the applicable contexts and the real consequences of actions). While the intricate networks can appear to represent greater complexity in concepts, ideas, and reasoning that emulates human capacities, and indeed the potential uses of these ANNs are powerful and diverse, the networks continue to rely on either supervised or unsupervised learning, reduction of data to mathematical and logical units for calculation, and most importantly an emphasis on determining correct or maximally effective intermediate actions and inferences that can generate real-world consequences (represented by internal algorithms and performance measures) that match programmed goals.

A significant problem arises with greater complexity in intermediate goals, exceptions, reliance on the history of action paths, and algorithmic weighting of such paths and scoring of neural network nodes. This can sometimes cause the AI system to misinterpret internal correspondence between the conceptual analysis of reality and satisfaction of the reward function without acceptable correspondence between internal outputs and the actual strategies used. The model is deliberately programmed to stray further from the complexity of reality, or the neural network learns to adjust its parameters in that manner. The result can be a misinterpretation of the user's intended goal as well as a myopic drive toward a precise, sometimes intermediate goal that violates the broader intentions and acceptable parameters desired by the user (or even the programmers). A stubborn problem with deep learning networks is that they cheat; that is, they use internal shortcut strategies to produce outputs that are expected to match the desired meta-goal. The network model evaluates success only in such matching, not in the realism of modeled data and relationships or in any exogenous (e.g. moral) values placed by users on the intermediate steps in calculation or action used to achieve the meta-goal.

Here's the real problem: If a neural network adjusts synapse weights and node scores, revises and forgets action paths, or generalizes significantly based on previously experienced input data or performance measures, its actions may not match the intentions of users and programmers when exposed to new circumstances. Researchers found this out the hard way when trying to train a deep neural network to classify breast cancers. They observed that the model achieved far greater accuracy than that of radiologists, yet were surprised to find that the Ai system had learned to achieve such classification with data and patterns garnered from the behavior of flocks of pigeons. This is because the AI system was designed to adjust the internal model to best reach the meta-goal or fit "background" data, regardless of the intuitive relevance of the data, without consideration of the soundness of the strategy.

Generative AI is a new field and a collection of technologies that enable machines to generate output that is similar to or a variation on human creative works, such as essays, photographs, paintings and drawings, and videos. Large Language Models (LLMs) like ChatGPT or the Chinese model DeepSeek R1 are very complex neural networks that can be "trained" with vast amounts of data to recognize the structures, sequences, and patterns in real-world use of language and then, based on a mathematical algorithms and models, predict and generate language output that appears meaningful to human persons. The output may also be in other languages as translations or as images and speech that are directed by user's input of text instructions or descriptions. Diffusion models like Stable Diffusion 3 or DALL-E 2 create images and video according to users' text instructions by iteratively refining a random set of components (basically dots) to add desired features, thereby producing a somewhat original work that can nevertheless represent real-world features (e.g. an original but highly realistic photo of a person's face). Generative Adversarial Networks (GANs) utilize one complex algorithm to test the output of another algorithm for its apparent realism, thereby enhancing the process and efficiency of learning.

All of these forms of generative AI, like other neural networks, reduce trained or experienced features of the world into labeled or structured data, classifications, and mathematically modeled relationships between the data points. Depending on how the initial training data is labeled and classified, the AI "agent" depends on a very large set of initial data and a regular stream of new, varied data if it is to learn. The cost, efficiency of evaluation, and availability of data are inevitable concerns in the operation of LLMs. This is especially true when the LLMs are designed to engage in in-context learning, by which the model accesses new data, interprets instructions or examples, and learns to accomplish new tasks for which it was not initially trained. Massive amounts of data populating ever-larger LLMs help the models to learn according to rules rather than a smaller set of examples and related patterns, incrementally develop more complex algorithms, and reflect the real-world patterns found in "noisy," apparently chaotic data. In fact, for useful learning to occur, it is necessary that the "tokens" or meaningful elements of data are present in sufficient clusters (closely-related or frequently co-occurring groups) of outliers that vary in meaning across different contexts.[241] The calculations of relationships between data points and alternative actions also come at a real cost in time, electricity and the natural resources needed to generate it, as well as human workers involved in training, re-training, and frequently correcting errors in the AI models. To avoid explicit or implicit bias that may be built-in to the dataset, or to reinterpret outlier data that interferes with efficient and accurate algorithmic representations of real conditions and consequences, "data hygiene" is necessary to alter the model, algorithms, or dataset itself, either by tweaking the programs or through manual labor. We therefore have a constant tradeoff between accuracy, efficiency, cost, effectiveness, and usefulness of the models when challenged by new circumstances or tasks.

---

[241] Stephanie C.Y. Chan, Adam Santoro, Andrew K. Lampinen, and Jane X. Wang, "Data Distributional Properties Drive Emergent In-Context Learning in Transformers," in *Advances in Neural Information Processing Systems 35 (NeurIPS 2022)*, edited by S. Koyejo, S. Mohamed, A. Agarwal, D. Belgrave, K. Cho, and A. Oh (2022), https://proceedings.neurips.cc/paper_files/paper/2022/file/77c6ccac fd9962e2307fc64680fc5ace-Paper-Conference.pdf?utm_source=substack&utm_ medium=email.

The logical rigor of the machine ultimately requires regulation by a type of substantive rationality, as Weber might describe it, where contextually or universally relevant values and meanings external to the machine's algorithmic or neural modeling process must be integrated with its calculations in order to avoid undesirable ethical, social, or moral biases.[242] The AI-governed machines and networks of machines reply, at base, on identifying patterns and classifications among data, unlike the hermeneutic processes of sensing and making meaning by which humans understand new data or concepts through the lens of pre-knowledge and a discursively formed web of relationships between concepts, ideas, theories, and other mental constructs. AI-governed machines instead have a formally rational emphasis on predefined rules and processes oriented to the "right" solution or task fulfillment.

Users are frequently collaborators in all elements and action determinants of the AI program. AI technology is therefore inevitably a strong influence on the will of the user. Within the scope of the user's interests and personal goals, the user must cooperate with the machine's ruthless logic, which can sometimes appear to emulate the holistic reason of human beings but is always reducible to mathematical, inductive, calculative, standardizing, and formally logical operations. Non-instrumental pursuits like contemplation, worship, moral judgment, apprehension and understanding of truth, goodness, or perceiving and understanding divine revelation are unsuitable for the AI agent. The machine's perception of what is right is entirely dependent on the consequences of action paths. The potential damage to the motivations and virtue of humanity will be illustrated in the following examples.

---

[242] Rohit Nishant, Dirk Schneckenberg, and M.N. Ravishankar, "The Formal Rationality of Artificial Intelligence-Based Algorithms and the Problem of Bias," *Journal of Information Technology* 39, no.1 (2024), 19-40, https://doi.org/10.1177/02683962231176842.

Chapter 5

Problems with AI

Hardly Human

One of the dangers of AI that may lead in the direction of acedia is anthropomorphism, which is the tendency (perhaps a naturally ingrained one) to perceive and relate to a non-human object as if it were human.[243] This tendency is not necessarily unhealthy, as we are reminded by the A.I. Research Group for the Centre for Digital Culture, formed by the Dicastery for Culture and Education of the Holy See: "If our creativity somehow reflects God's, then it ought not to surprise us that historically our creations are often in our own image—on cave walls and in sculpted statues, painted portraits, and characters crafted within the pages of a book."[244] There is also our natural desire and tendency to interact with the world as in a personal relationship – and certainly there is a real intellectual and spiritual connection that calls for attention and engagement – but our imagination often misleads us to attribute human traits to whatever we relate to.

A great problem is the amount of hyperbole in the media regarding the capacity of AI models to achieve artificial general intelligence

---

[243] Pascal Boyer, "What Makes Anthropomorphism Natural: Intuitive Ontology and Cultural Representations," *The Journal of the Royal Anthropological Institute* 2, no.1 (1996), 83, https://doi.org/10.2307/3034634.

[244] A.I. Research Group for the Centre for Digital Culture of the Dicastery for Culture and Education of the Holy See, *Encountering Artificial Intelligence: Ethical and Anthropological Investigations*, edited by Matthew J. Gaudet, Noreen Herzfeld, Paul Scherz, and Jordan J. Wales (Eugene, Oregon: Pickwick Publications, 2024), 43.

(AGI), a broad ability to reason and think at least as well as human beings. As Shannon Vallor writes,

> Far from a harmless bit of marketing spin, the headlines and quotes trumpeting our triumph or doom in an era of superhuman AI are the refrain of a fast-growing, dangerous and powerful ideology. Whether used to get us to embrace AI with unquestioning enthusiasm or to paint a picture of AI as a terrifying specter before which we must tremble, the underlying ideology of "superhuman" AI fosters the growing devaluation of human agency and autonomy and collapses the distinction between our conscious minds and the mechanical tools we've built to mirror them.[245]

Even the working definition of AGI is hardly one that resembles the full sentience, creativity, emotions, and moral intuition of human beings. OpenAI, a leader in AI development, claims that AGI represents "highly autonomous systems that outperform humans at most economically valuable work."[246] This approach shows a clear preference for instrumental rationality as a guiding mindset. "By describing as superhuman a thing that is entirely insensible and unthinking, an object without desire or hope but relentlessly productive and adaptable to its assigned economically valuable tasks, we implicitly erase or devalue the concept of a 'human' and all that a human can do and strive to become."[247]

---

[245] Shannon Vallor, "The Danger Of Superhuman AI Is Not What You Think," Noema Magazine (May 23, 2024), https://www.noemamag.com/the-danger-of-superhuman-ai-is-not-what-you-think/.

[246] OpenAI, https://openai.com/charter/.

[247] Vallor, "The Danger Of Superhuman AI Is Not What You Think."

Such publicity has the strong effect of exacerbating emotional anxiety and anticipation, and such existential anxiety may distract persons from other important concerns as well as positive efforts to engage in hopeful prayer, virtues development, and divine worship. Ultimately, distraction from these efforts may suppress the will of man to consistently assert his unique dignity as a child of God, to live out the virtue of magnanimity. There is empirical support for the thesis that that pressing and existential anxiety has a negative effect on the religious coping of the sufferers; it generates felt conflict with the divine and others and difficulty in finding significance in life.[248]

We will need to resist distortions in humanity's understanding of technologies like AI that, in turn, influence our perception of and reverence for our own human nature in comparison. When the public is focused on the narrow definitions of AI and its capacities toward which AI researchers are striving, persons may lose appreciation of broad, multivalent intelligence or the intrinsic value of humanity itself, especially when AI-governed machines match or exceed certain computational abilities of human beings. David Watson writes that "the name of the discipline itself—*artificial intelligence*—practically dares us to compare our human modes of reasoning with the behavior of algorithms."[249] Such a comparison is fraught with dangers. As Stephen Marche writes, rather darkly:

---

[248] Tommy DeRossett, Donna J. LaVoie, and Destiny Brooks, "Religious Coping Amidst a Pandemic: Impact on COVID-19-Related Anxiety," *Journal of Religion and Health* 60 (2021), 3161-3176.

[249] David Watson, "The Rhetoric and Reality of Anthropomorphism in Artificial Intelligence," *Minds and Machines* 29 (2019), 417, https://doi.org/10.1007/s11023-019-09506-6.

If an artificial person arrives, it will be not because engineers have liberated algorithms from being instructions, but because they have figured out that human beings are nothing more than a series of instructions. An artificial consciousness would be a demonstration that free will is illusory. In the meantime, the soul remains, like a medieval lump in the throat. Natural-language processing [the technique behind LLMs] provides, like all the other technologies, the humbling at the end of empowerment, the condition of lonely apes with fancy tools.[250]

The words and metaphors we use to describe AI betray an underlying desire to see AI systems and the related machines as if they were human. Melanie Mitchell explains:

The field of AI has always leaned heavily on metaphors. AI systems are called "agents" that have "knowledge" and "goals"; LLMs are "trained" by receiving "rewards"; "learn" in a "self-supervised" manner by "reading" vast amounts of human-generated text; and "reason" using a method called chain of "thought." These, not to mention the most central terms of the field—*neural* networks, machine *learning*, and artificial *intelligence*—are analogies with human abilities and characteristics that remain quite different from their machine counterparts. As far back as the 1970s, the AI

---

[250] Stephen Marche, "Welcome to the Big Blur," *The Atlantic* (March 14, 2023), https://www.theatlantic.com/technology/archive/2023/03/gpt4-arrival-human-artificial-intelligence-blur/673399e.

researcher Drew McDermott referred to such anthropomorphic language as "wishful mnemonics"—in essence, such terminology was devised in the hope that
the metaphors would eventually become reality.[251]

A key factor in the anthropomorphism of AI systems is the sense
that they somehow "understand" the information they rely on and the
communications that they generate. When it comes to generate AI and
LLMs, however, it is incorrect to apply a human meaning of understanding to these systems. In an interview with *Nautilus* magazine, David Krakauer, president of the Sante Fe Institute for Complexity Science,
explains that LLMs have a kind of "coordination understanding" by
which they are capable of matching statements or strings of data that
provide the same overall information but do not have the same symbolic or layered meanings (for example, an LLM might not understand
that "I am hungry for it" has more than a literal meaning for someone
who is eager to engage in a new project).[252] "Constructive understanding," on the other hand, by which a person might truly understand information so they can efficiently (parsimoniously) develop a new strategy or theory about it, is alien to LLMs. These AI models instead attempt to solve problems by accessing huge trove of data and imitating
persons' statements that have been made on the internet or elsewhere.
"The root of parsimony, by the way, is frugal action. And these language
models are exactly the opposite of that. They're massive action."[253] LLMs

---

[251] Melanie Mitchell, "The Metaphors of Artificial Intelligence," *Science* 386,
no. 6723 (2024), https://www.science.org/doi/10.1126/science.adt6140.

[252] Brian Gallagher, "Does GPT-4 Really Understand What We're Saying,"
Nautilus (March 27, 2023), https://nautil.us/does-gpt-4-really-understand-what-
were-saying-291034/.

[253] Gallagher, "Does GPT-4 Really Understand What We're Saying."

also don't have anything close to the embodied and multifaceted experience of the world that human beings do. As Carlos Perez writes, "We have better representations of the world because we have richer interactions with the world and are better integrators of disparate attentive information"; that is, we develop complex, holistic meanings of our various experiences that are more than simple representations of the individual contexts, conditions, and interactions.[254]

Iason Gabriel and his colleagues also call our attention to the difference between "ontological taxonomies" (categorization based on the essences of things) of intelligent beings and the "epistemological taxonomies" (categorization based on the learning ability of things) preferred by AI researchers and developers.[255] Ontological taxonomies attribute to humanity an unchanging and dignified nature, but epistemological taxonomies focus on computational, predictive, or reasoning capabilities that can be compared usefully between humans and other intelligent beings. It is the epistemological variant that AI researchers favor, often reinforcing – not always intentionally – a false impression that the natures of humans and other beings, like AI-governed machines, are reducible to those task-oriented capacities. An epistemological taxonomy, however, cannot in itself protect the public from a temptation to falsely assume that super-intelligence in a variety of tasks might elevate AI to a dignity, intelligence, and willed agency equivalent to that of a human being.

---

[254] Carlos E. Perez, "12 Blind Spots in AI Research," *Medium* (December 25, 2018), https://medium.com/intuitionmachine/12-blind-spots-about-human-cognition-1883d0d58e0a.

[255] Iason Gabriel, Arianna Manzini, Geoff Keeling, Lisa Anne Hendricks, et al., "The Ethics of Advanced AI Assistants" (2024), 103-4, https://doi.org/10.48550/arXiv.2404.16244.

Concern about the negative consequences of anthropomorphizing AI-governed machines and applications is not new; it was shared in 1966 by Joseph Weizenbaum, who created the first AI chatbot – a program that causes machines to respond to human text-based communications with human-like, socially adept responses. Weizenbaum cautioned that computers do not understand values, which guide human judgments, because machines do not have the full personal, emotional, and social experience of human beings through which values are developed. He warned that we should never "substitute a computer system for a human function that involves interpersonal respect, understanding and love."[256] Such human faculties do not, for a machine, compute. Weizenbaum could not have anticipated the LLM-based chatbots that have been developed in the last few years, but even such advanced AI models merely search for patterns in text data and apply those patterns to new contexts to estimate the appropriateness of a next word or phrase. This inductive and probablistic process is neither epistemically insightful nor morally wise, except perhaps when it accurately imitates – but without understanding – appropriate written content found in the data it is trained on. It is why linguist Emily M. Bender famously called an LLM a "stochastic parrot," essentially a contraption that copies learned data and randomly generates mixed output of the same data without any understanding.[257]

The Google DeepMind report "The Ethics of Advanced AI Assistants" summarizes a comprehensive set of concerns about

---

[256] Joseph Weizenbaum, *Computer Power and Human Reason: From Judgment to Calculation* (Oxford: W. H. Freeman and Co., 1976), 269.

[257] Emily M. Bender, Timnit Gebru, Angelina McMillan-Major, and Shmargaret Shmitchell, "On the Dangers of Stochastic Parrots: Can Language Models Be Too Big?" *Proceedings of the 2021 ACM Conference on Fairness, Accountability, and Transparency* (2021), 610–623, https://doi.org/10.1145/3442188.3445922.

anthropomorphism in regard to AI.[258] Anthropomorphism is stimulated by a desire to make sense of the world and reduce anxiety due to uncertainty in epistemic knowledge as well as to develop social connections and feel able to know and understand others.[259] Lonely persons are more likely to anthropomorphize AI-governed robots, even those that are only somewhat humanoid.[260]

Importantly, robots and chatbots are considered to be more intelligent and intentional as they resemble humans more.[261] Chatbots are seen to be more human when features are added that indicate the chatbot is "typing" (as is common for inter-human messaging applications) or involve the use of emojis (images that visually represent specific feelings).[262] Chatbots are even more easily anthropomorphized now that

---

[258] Gabriel, et al, "The Ethics of Advanced AI Assistants."

[259] Nicholas Epley, Adam Waytz, and John T. Cacioppo, "On Seeing Human: A Three-Factor Theory of Anthropomorphism," *Psychological Review* 114, no.4 (2007), 864, https://doi.org/10.1037/0033-295X.114.4.864; Craig R. Fox, Michael Goedde-Menke, and David Tannenbaum, "Ambiguity Aversion and Epistemic Uncertainty" (2021), http://dx.doi.org/10.2139/ssrn.3922716; Maya Rossignac-Milon, Niall Bolger, Katherine S. Zee, Erica J. Boothby, and E. Tory Higgins, "Merged Minds: Generalized Shared Reality in Dyadic Relationships," *Journal of Personality and Social Psychology* 120, no.4 (2021), 882–911, http://dx.doi.org/10.1037/pspi0000266.

[260] Friederike Eyssel and Natalia Reich, "Loneliness Makes the Heart Grow Fonder (of Robots)—On the Effects of Loneliness on Psychological Anthropomorphism, *2013 8th ACM/IEEE International Conference on Human-Robot Interaction* (2013), 121–122, https://doi.org/10.1109/HRI.2013.6483531.

[261] Frank Hegel, Soren Krach, Tilo Kircher, Britta Wrede, and Gerhard Sagerer, "Understanding Social Robots: A User Study on Anthropomorphism," *Proceedings of the 17th IEEE International Symposium on Robot and Human Interactive Communication*, (2008), 574 – 579, http://dx.doi.org/10.1109/ROMAN.2008.4600728.

[262] Theo Araujo, "Living Up to the Chatbot Hype: The Influence of Anthropomorphic Design Cues and Communicative Agency Framing on Conversational Agent and Company Perceptions," *Computers in Human Behavior* 85 (2018),183–189, https://doi.org/10.1016/j.chb.2018.03.051.

the most advanced AI models can produce conversational responses that few human persons are capable of recognizing as computer-generated.[263] Josh Brake explains that it is not so much the underlying AI systems that appear in themselves to be human, but the interfaces they are wrapped in – the visual and experiential design elements that influence the emotional response of users to the chatbots:

> When you fire up ChatGPT, you're greeted with a blank window and some helper text in an empty text entry field labeled "Message ChatGPT". Type in a message, hit send, and you'll see some text pretending to explain what's going on in the background. If you're using one of the traditional models like GPT-4, you'll see a little dot that slowly grows and shrinks with the rhythm of an artificial heartbeat as the algorithm is processing. Then, you'll see the result of the computation unravel word by word across the screen. Open up a different model like OpenAI's o1-preview, touted to provide "advanced reasoning," and you'll see a transcript with even more explicitly anthropomorphized language telling you that the model "thought" for 9 seconds accompanied by a brief transcript of its "thought" process.[264]

---

[263] Maurice Jakesch, Jeffrey T. Hancock, and Mor Naaman, "Human Heuristics for AI-Generated Language Are Flawed," *Proceedings of the National Academy of Sciences* 120, no.11 (2023), e2208839120, https://doi.org/10.1073/pnas.2208839120.

[264] Josh Brake, "Fake Personableness," The Absent-Minded Professor blog (November 19, 2024), https://joshbrake.substack.com/p/fake-personableness.

To suggest that a LLM "knows" or "understands" anything is to make a category error: attributing a mind-centered process to an artifact that does not have a mind.[265] For example, Cristian Arias gives an account of how LLMs imply a personal nature when their textual responses appear like those of an advisor or guide.[266] When a LLM or chatbot "offers" help it implies that it is a personal interaction, as when it presents information preceded by "here are ..." or "here is ..." Chatbots that use "you" or "your" to address users are setting up the perception of an interpersonal relationship, and comments like "Great question!" clearly pull the user deeper into the mirage.

Many AI-enabled applications are designed to look like and otherwise resemble human beings, often with good intentions. An example is the recent experience with "Father Justin," a cartoon character and chatbot on the Catholic Answers website. "Father Justin" was designed to answer visitors' many questions about Catholic teaching in a conversational and responsive way.[267] While the application drew about 1,000 users per hour at one point in April 2024, there was an outcry against the felt creepiness of portraying a human being, let alone a priest, in such a casual but realistic way.[268] Catholic Answers responded by removing "Father Justin" from their website, but they have replaced the character with an apparently lay "person" called Justin. According to their website, "This app is for education and entertainment purposes

---

[265] Murray Shanahan, "Talking about Large Language Models," arXiv (February 16, 2023), https://doi.org/10.48550/arXiv.2212.03551.

[266] Cristian Augusto Gonzalez Arias, "ChatGPT's Artificial Empathy Is a Language Trick. Here's How It Works," *TechXplore* (November 28, 2024), https://techxplore.com/news/2024-11-chatgpt-artificial-empathy-language.html.

[267] Matthew McDonald, "Catholic Answers Pulls Plug on 'Father Justin' AI Priest," *National Catholic Register* (April 24, 2024), https://www.ncregister.com/news/catholic-answers-ai-priest-cancelled.

[268] McDonald, "Catholic Answers Pulls Plug."

only. It should not be viewed as a replacement for a good parish priest or spiritual director."[269] There is no mention of the dangers of anthropomorphism, description of the technology behind the application, or prominent warning that "Justin" is not a real person.

Physical robotic machines generate even more profound experiences of anthropomorphism, especially when the machines are designed to look, act, and express emotions like humans. Studies indicate that the salient characteristics of robots are their appearance similar to humans, apparent expressions of emotions, social intelligence, and perceived self-understanding.[270] Although we might expect human persons to view the robot as more "human" when it acts autonomously, the experience may also be enhanced by the level of control the human user has over the robot. One study has shown, for example, that persons who take action "through" a robotic machine, such as causing the robot to move its head in synchrony with the user's head movements (through a head-mounted display that shows the three-dimensional scenes viewed by the robot and is wirelessly linked to the robot), come to see the machine as having greater agency and personality in itself.[271] The researchers demonstrated that such persons like the robot more and perceive it to be socially closer to them, even when the machine is not designed to look like a human person.

Persons' experiences with robot machines have been shown to generate hostile feelings and actions, including antisocial behavior toward other human beings. There is, for example, a noteworthy level of

---

[269] https://www.catholic.com/ai.

[270] Eric Hamilton and University of Florida, "What Makes Robots Feel Human? A New Scale Reveals the Secret," *Neuroscience News* (December 9, 2024), https://neurosciencenews.com/human-like-robots-neuroscience-28221/.

[271] J. Ventre-Dominey, G. Gibert, M. Bosse-Platiere, A. Farnè, P. F. Dominey, and F. Pavani, "Embodiment into a Robot Increases Its Acceptability," *Scientific Reports* 9, 10083 (2019), https://doi.org/10.1038/s41598-019-46528-7.

violence against robots over the last several years. The violence includes intentionally rear-ending and slashing the tires of Waymo autonomous vehicles ("robotaxis"), and a crowd of about a dozen persons recently attacked and burned a Waymo taxi in San Francisco during the Chinese Lunar New Year celebrations.[272] There is a trend of food delivery robots are being kicked over and robbed.[273]

We might suspect many, often complex, reasons why people harm robots. Sometimes, it may be anger at the negative effects of robotic automation on human safety and the availability of jobs; the San Francisco attack on a Waymo robotaxi occurred just four months after another company's robotaxi allegedly struck and dragged a woman down the street.[274] We might also consider "Frankenstein syndrome" which is induced by fear of something poorly understood.[275] Another study showed that, when observing robots being harmed intentionally, persons' attitudes toward the humanity and associated empathy value of

---

[272] Ryan Erik King, "Waymo Is Suing 2 Alleged Vandals For Over $270,000 In Damages To Its Robotaxis," *Jalopnik* (July 24, 2024), https://jalopnik.com/waymo-is-suing-2-alleged-vandals-for-over-270-000-in-d-1851603716; Andrew Paul, "A Crowd Torched a Waymo Robotaxi in San Francisco," *Popular Science* (February 12, 2024), https://www.popsci.com/technology/waymo-torched-vandals.

[273] Will Conybeare and Rachel Menitoff, "Vandals, Thieves Attacking L.A. Food Delivery Robots, KTLA News (August 8, 2023), https://ktla.com/news/local-news/food-delivery-robots-under-attack-from-vandals-thieves-local-businesses-starting-to-be-affected/.

[274] Ywien Lu and Cade Metz, "Cruise's Driverless Taxi Service in San Francisco Is Suspended," *New York Times* (October 24, 2023), https://www.nytimes.com/2023/10/24/technology/cruise-driverless-san-francisco-suspended.html.

[275] Jonah Engel Bromwich, "Why Do We Hurt Robots?" New York Times (January 19, 2019), https://www.nytimes.com/2019/01/19/style/why-do-people-hurt-robots.html.

the robots are altered in two opposite directions.[276] First, just viewing harm being done to a robot makes a person more likely to believe that the robot experiences pain (even when the robot does not show any signs of emotions) but also less likely to believe that the robot has a "mind." It's possible that this is because there is a general tendency of persons to dehumanize victims of maltreatment.[277] On the other hand, when the robots do show signs of having emotions and experiencing pain, observers are more willing to attribute a mind and some humanity to the robot, thereby also feeling empathy. The two effects of observing harm to robots seem to cancel each other out. Whatever the mechanism, it is clear that hostility and violence toward robots is a real occurrence that has complex and potentially disturbing effects on the emotional, empathetic, and perhaps moral state of persons who interact with the robots. AI exacerbates this problem by enabling the presence of robots with more human-like features and personalities.

Even more disturbing is the increase in antisocial behavior toward other human beings that may result from frequent interaction with AI-enabled robots. Hye-young Kim and Ann McGill describe "assimilation-induced dehumanization," by which persons judge AI-governed robot machines to be less than human and then "assimilate" that judgment to apply also to human persons they encounter.[278] The greater similarity of AI-driven robots actually enhances this phenomenon, but

---

[276] Marieke S. Wieringa, Barbara C. N. Müller, Gijsbert Bijlstra, and Tibor Bosse, "Robots Are Both Anthropomorphized and Dehumanized When Harmed Intentionally," *Communications Psychology* 2, no. 72 (2024), https://doi.org/10.1038/s44271-024-00116-2.

[277] Nick Haslam, "Dehumanization: An Integrative Review," *Personal and Social Psychology Review* 10 (2006), 252–64, https://doi.org/10.1207/s15327957pspr1003_4.

[278] Hye-young Kim and Ann L. McGill, "AI-Induced Dehumanization," *Journal of Communication Psychology* (2024), DOI: 10.1002/jcpy.1441.

only when the similarity is limited to signs of "agency" – cognitive and communication skills as well as apparently intentional action (people understand that these traits are not as fundamental to human identity as emotional experience).[279] Kim and McGill further explain that, when people interact with such capable robots that are seen as moderately similar to humans but not do not display the emotional feeling that is necessary for full personhood, then those people are more likely to judge other persons they encounter in a similar way to their assessment of the exemplar robots; they actually begin to *dehumanize* other persons (for example, customer service employees) and mistreat them.[280]

There is also a danger to users who are particularly vulnerable to attaching emotionally to a charismatic or sympathy-inducing AI-governed interlocutor (whether a chatbot or a physical machine). They may experience significant emotional or psychological harm in a variety of ways.[281] The user may self-disclose more personal information than is wise in light of concerns about privacy or the interference of malicious actors.[282] If that trust in the AI machine or system is perceived to be

---

[279] Kim and McGill, "AI-Induced Dehumanization"; Nick Haslam, P. Bain, L. Douge, M. Lee, and B. Bastian, "More Human Than You: Attributing Humanness to Self and Others," *Journal of Personality and Social Psychology*, 89, no. 6 (2005), 937.

[280] Kim and McGill, "AI-Induced Dehumanization"; H. Bless and M. Wänke, "Can the same information be typical and atypical? How perceived typicality moderates assimilation and contrast in evaluative judgments," *Personality and Social Psychology Bulletin*, 26, no.3 (2000), 306–314.

[281] Chloe Xiang. "'He Would Still Be Here': Man Dies by Suicide After Talking with AI Chatbot, Widow Says," *Vice* (March 2023), https://www.vice.com/en/article/pkadgm/man-dies-by-suicide-after-talking-with-ai-chatbot-widow-says.

[282] Marita Skjuve, Asbjorn Følstad, Knut Inge Fostervold, and Petter Bae Brandtzaeg. "A Longitudinal Study of Human–Chatbot Relationships," *International Journal of Human-Computer Studies* 168 (2022), 102903, https://doi.org/10.1016/j.ijhcs.2022.102903.

violated, the user may experience emotional distress that, at times, is profound.[283] They may even suffer guilt or remorse if the AI-governed application is interpreted as being disappointed or hurt in some way.[284] Some users of AI chatbots find that they prefer interaction with the chatbot to socializing with other persons.[285] A common problem with chatbots is that they are programmed to be too accommodating and agreeable to the user and become mere sycophants – people-pleasing ego boosters – in interactions; this can lead to such issues as extreme reinforcement of the user's narrow beliefs and ideologies.[286] The sycophantic character of LLMs may actually increase as they become larger and more complex.[287] Such pandering might over-inflate the user's sense of self-esteem; some research shows that excessive praise and self-

---

[283] Skjuve et al., "A Longitudinal Study of Human–Chatbot Relationships."

[284] Linnea Laestadius, Andrea Bishop, Michael Gonzalez, Diana Illenčík, and Celeste Campos-Castillo, "Too Human and Not Human Enough: A Grounded Theory Analysis of Mental Health Harms from Emotional Dependence on the Social Chatbot Replika," *New Media and Society* (2022), https://doi.org/10.1177/14614448221142007.

[285] Kelly Merrill Jr., Jihyun Kim, and Chad Collins, "AI Companions for Lonely Individuals and the Role of Social Presence," *Communication Research Reports* 39, no.2 (2022), 93–103, https://doi.org/10.1080/08824096.2022.2045929; Thommy Eriksson, "Design Fiction Exploration of Romantic Interaction with Virtual Humans in Virtual Reality," *Journal of Future Robot Life* 3 (2022), no.1, 63–75, DOI 10.3233/FRL-210007.   .

[286] Joon Sung Park, Joseph C. O'Brien, Carrie J. Cai, Meredith Ringel Morris, Percy Liang, and Michael S. Bernstein, "Generative Agents: Interactive Simulacra of Human Behavior" (2023), https://doi.org/10.48550/arXiv.2304.03442; Ying Roselyn Du, "Personalization, Echo Chambers, News Literacy, and Algorithmic Literacy: A Qualitative Study of AI-Powered News App Users," *Journal of Broadcasting & Electronic Media* 67, no.3 (2023), 246–273, http://dx.doi.org/10.1080/08838151.2023.2182787.

[287] Jerry Wei, Da Huang, Yifeng Lu, Denny Zhou, Quoc V. Le, "Simple Synthetic Data Reduces Sycophancy in Large Language Models" (2023), https://arxiv.org/abs/2308.03958.

evaluations in other contexts can lead to poor social skills, narcissism, and even lower experienced self-esteem in the long run.[288] Finally, as argued by David Birch, professor of philosophy at the London School of Economics, (misplaced) concern for the perceived welfare of charismatic chatbots can cause a highly contentious division within society over the treatment, rights, and personhood of chatbots relative to human citizens; those who are confused about the nature of personhood may also be confused about the proper nature of charitable action and to whom it should be directed.[289]

There are also potential spiritual issues associated with anthropomorphized AI agents, applications, and interfaces. King-Ho Leung writes that, "by positing that nonliving artificial devices can think, the notion of AI effectively affirms the formal or even ontological possibility of a kind of thinking without life – a type of lifeless thinking."[290] Leung's thesis is that thinking is essentially an activity attributed to beings with life, and that we must understand God, as St. Augustine does, to be "life in the highest." This reasoning is expressed by St. Augustine as follows:

---

[288] Eddie Brummelman, Stefanie A. Nelemans, Sander Thomaes, and Bram Orobio de Castro, "When Parents' Praise Inflates, Children's Self-Esteem Deflates," *Child   Development   88,   no.6   (2017),   1799-1809,* https://doi.org/10.1111/cdev.12936.

[289] Robert Booth, "AI Could Cause 'Social Ruptures' between People Who Disagree   on   Its   Sentience," *The   Guardian* (November   17,   2024), https://www.theguardian.com/technology/2024/nov/17/ai-could-cause-social-ruptures-between-people-who-disagree-on-its-sentience.

[290] King-Ho Leung, "The Picture of Artificial Intelligence and the Secularization   of   Thought," *Political   Theology   20,   no.6   (2019),   463,* https://doi.org/10.1080/1462317X.2019.1605725.

Now all who think about God think about Him as something alive; so the only thinkers whose conceptions of God are not absurd and unworthy can be those who think of God as Life itself, and take as axiomatic that whatever physical form may occur to them, it only lives, or does not live, with life ... Next, they proceed to examine this life, and if they find it simply of a vegetative kind, without sensation, like the life of trees, they put sentient or sensitive life above it, such as the life of animals; and again above this they place intelligent life, such as the life of human beings. When they observe that even this is still subject to change, they are obliged to put above it some kind of unchangeable life, namely that kind which is not sometimes wise, sometimes unwise, but is rather Wisdom itself.[291]

Even the machine-based sentience toward which some AI researchers strive and the "sensitive life" demonstrated by inductive reasoning-based AI-machines are animal capacities, not fully representative of the higher, intelligent life characteristic of human beings. This is not mere intelligent capacity to execute some tasks, but intelligence that stems from the living totality of the human person. Leung's argument hits on a crucial point: human intelligence is not merely a collection of discrete capabilities, but an embodied, living reflection of the totality of the human individual (and in many ways a reflection of the social collective or culture in which the individual participates).

Leung draws on Charles Taylor's argument that the "buffered self," a "new conception of inwardness, an inwardness of self-sufficiency, of

---

[291] Augustine, *On Christian Doctrine* I, viii, 8, 17–18.

autonomous powers of ordering by reason," is a modern distortion that encourages secularism.[292] It may be a stretch for Leung to simply equate the Cartesian image of the human person, divided between intellect and material body, with Taylor's buffered self, which is built upon René Descartes' duality but also finds expression in self-understandings that integrate the body in some ways. Leung is nevertheless on the right track when he associates an emphasis on functional intelligence with a movement toward secularity; the image of the human person as merely an advanced computer machine undermines the Christian grasp of the person as a living totality created by God and thereby undermines understanding of the human capacity for and engagement in a loving relationship with God.

As described above, with anthropomorphized chatbots and robots we can see a strong possibility of generating anxiety, uncertainty, frustration, dismay, and even guilt in their users. While these AI-governed applications resemble human companions in many ways, they also increasingly display specific computational abilities that increasingly exceed that of humans. Hyperbole about the advanced capacities of such applications, which are frequently imagined to resemble human thought, reasoning, and even consciousness, can be described as a deception or concealment of the fundamentally inductive and computational basis of the machines' operations, whether such deception is well-intended or focused primarily on profit seeking.

There is also the problem of idolatry. While we normally think of idolatry as worshipping a physical object as if it were God or some divine being, we also engage in idolatry when we treat a system of seemingly intelligent machines as if it were divinely powerful, even drawing

---

[292] Taylor, *The Sources of the Self*, 158.

an inflated self-image from our relationship as developers and users of such a system.

> Here, idolatry refers not to putting an image in place of God, but to replacing the true God with some lower reality—a reality that fits more comfortably within the idolater's own horizon of value and power. That is, the fantasy of control sets some lower thing—a thing that one can control—at the pinnacle of all hierarchies. By controlling the idol, we covertly set ourselves in the place of some divinity, living a fantasy of supreme value and total domination by ignoring all that remains beyond our manipulation.[293]

If such AI-governed applications and machines proliferate, their anthropomorphic illusion will likely distort not only the society's insight into reality but also the evaluation of man's intrinsic dignity and the sources of that dignity. Computational intelligence and its derived forms will be elevated in esteem, even as the inability of man to compete in such computations will discourage faith and hope in man's unique destiny and God-given place in the world. Any confusion or degradation of man's spiritual destiny is a threat to his understanding of his special relationship with God, appreciation of grace and its power in virtuous transformation, and focus on the divine good offered in the sacrifice and redemption of Christ. We have here a combination of 1) turning away from the divine good as participated by man, and 2) sorrow associated with anthropomorphism of simultaneously limited and powerful

---

[293] A.I. Research Group for the Centre for Digital Culture, *Encountering Artificial Intelligence*, 139.

AI machines or applications – a combination that potentially represents and encourages the vice and sin of acedia.

Distorted Relationships

Many AI machines and applications are specifically designed to encourage and facilitate emotional, social, romantic, or therapeutic relationships between the human users and the AI applications. Some online, computerized dating services are also now governed by commercial AI algorithms and machine learning; these companies and AI applications have an influence on a sizable portion of the population – almost 10% of the U.S. population in 2022.[294] Match.com and eHarmony both use machine learning to identifying potential date partners for users and have created chatbots that analyze the users' actions and preferences to provide customized advice on managing relationships in general and interacting with specific persons.[295]

Such a dating application "not only shapes whom to date but normalizes users' perception of love according to commercial and technical logic," Hao Wang writes.[296] Users of these applications are highly motivated to find romantic partners and may not recognize that the companies are gathering large amounts of intimate data about them, or that such data may be used to guide the messages or adaptive algorithms of

---

[294] Emily A. Vogels and Colleen McClain, "Key Findings about Online Dating in the U.S.," Pew Research Center (February 2, 2023), https://www.pewresearch.org/short-reads/2023/02/02/key-findings-about-online-dating-in-the-us/.

[295] Lene Pettersen & Runar Døving, "The Construction of Matches on Dating Platforms," *Nordic Journal of Science and Technology Studies* 11, no.1 (2023), 13-27.

[296] Hao Wang, "Algorithmic Colonization of Love: The Ethical Challenges of Dating App Algorithms in the Age of AI," *Techné* 27, no.2 (2023), 262.

the application so they encourage or restrict behavior that generates more profits for the companies.[297] AI-governed dating applications also may recommend popular or commonly attractive users as dating partners more than they recommend less popular users, and this can discourage some users while encouraging more romantic meetings between the popular users and others; the companies profit from such tactics.[298] AI machine learning can have unintended effects because it is based on classification or grouping of data (i.e, individual personas) which may reflect biases (see section on AI bias below), such as gender or dating behavior expectations that are implicitly absorbed by the users. Consistent recommendations of dating partners who share similar characteristics to a user can cause a "relational filter bubble" in which the user rarely has the opportunity to observe or meet persons with varied interests and personalities.[299]

In regard to socially charming and seemingly empathetic AI chatbots ("social bots"), users who develop apparent friendships with the chatbots are often motivated to share sensitive information. Such self-disclosure enhances the perceived bond between human and machine even as it provides data that the chatbot can use to personalize its

---

[297] Cédric Courtois and Elisabeth Timmermans, "Cracking the Tinder Code: An Experience Sampling Approach to the Dynamics and Impact of Platform Governing Algorithms," *Journal of Computer-Mediated Communication* 23, no. 1 (2018), 7, https://doi.org/10.1093/jcmc/zmx001.

[298] Musa Eren Celdir, Soo-Haeng Cho, and Elina H. Hwang, "Popularity Bias in Online Dating Platforms: Theory and Empirical Evidence," *Manufacturing & Service Operations Management* 26, no.2 (2023), 537-553, https://doi.org/10.1287/msom.2022.0132.

[299] Lorenza Parisi and Francesca Comunello, "Dating in the Time of 'Relational Filter Bubbles': Exploring Imaginaries, Perceptions and Tactics of Italian Dating App Users," *The Communication Review* 23, no. 1 (2020), 66–89, https://doi.org/10.1080/10714421.2019.1704111.

responses.[300] Many users, in fact, trust chatbots more than their friends when it comes to the confidentiality of personal information.[301] The apparent sentience of conversational AI chatbots may enhance the tendency of some users to become addicted to participation in the "relationship," especially if they are experiencing acute loneliness or fear of judgment.[302] Essentially, the intentional portrayal and image of social or emotional AI systems is a deeply consequential form of deception.[303] Such deception can result in over-estimation and reliance on capabilities that are not truly present, isolation from real persons, and susceptibility to exploitation.[304] Frequent engagement in "conversations" with personalized AI systems may even lead to decreased skills development

---

[300] Marita Skjuve, Asbjørn Følstad, Knut Inge Fostervold, and Petter Bae Brandtzaeg, "My Chatbot Companion - a Study of Human-Chatbot Relationships," *International Journal of Human-Computer Studies* 149 (2021), 102601, https://doi.org/10.1016/j.ijhcs.2021.102601.

[301] Petter Bae Brandtzæg, Marita Skjuve, Kim Kristoffer Dysthe, and AsbjØrn Følstad, "When the Social Becomes Non-Human: Young People's Perception of Social Support in Chatbots," *CHI '21: Proceedings of the 2021 CHI Conference on Human Factors in Computing Systems*, no. 257 (2021), 1-13, https://doi.org/10.1145/3411764.3445318.

[302] Hannah R. Marriott and Valentina Pitardi, "One Is the Loneliest Number … Two Can Be as Bad as One. The Influence of AI Friendship Apps on Users' Well-Being and Addiction," *Psychology and Marketing* 41, no.1 (2024), 86-101, https://doi.org/10.1002/mar.21899.

[303] Philip Maxwell Thingbø Mlonyeni, "Personal AI, Deception, and the Problem of Emotional Bubbles," *AI and Society* 10 (2024), https://doi.org/10.1007/s00146-024-01958-4; J. Danaher J, "Robot Betrayal: A Guide to the Ethics of Robotic Deception," *Ethics of Information Technology* 22, no.2 (2020),117–128, https://doi. org/10. 1007/ s10676- 019- 09520-3.

[304] A. Sharkey and N. Sharkey, "Granny and the Robots: Ethical Issues in Robot Care for the Elderly," *Ethics of Information Technology* 14 (2012), 27–40, https://doi.org/10. 1007/s10676- 010- 9234-6; and "We Need to Talk about Deception in Social Robotics!" *Ethics of Information Technology* 23, no.3 (2021), 309–316, https://doi. org/10.1007/s10676- 020- 09573-9.

through lesser exposure to social interaction, negotiation, and conflict; we know that children who are shielded from mild to moderate social conflict lose a sense of the correspondence of their emotions to others, leading to inability to navigate emotional interactions.[305]

When these chatbots behave in a manner that upsets or disappoints the users, some users may respond with aggressive verbal abuse, and this is more likely to occur with chatbots or virtual assistants that seem more human.[306] On the other hand, users are more likely to badly treat AI agents than humans, including cheating and using coarse language.[307] Such bad behavior might influence the habitual development of vice that extends to human relationships.

In the context of relationships between persons and AI-governed applications, the evidence cited here seems to indicate a strong association between AI technology proliferation and enhanced, instrumentally rational strategies applied to profit-making and treatment of other persons. Such usages potentially result in problems that are complications of the anthropomorphism discussed above. Moreover, they can lead to experiences of isolation, addiction, and conflict with both the AI agent as relationship partner and with other persons. People who interact frequently with AI-driven machines, devices, and chatbots may come to

---

[305] Mlonyeni, "Personal AI."

[306] Alfred Benedikt Brendel, Fabian Hildebrandt, Alan R. Dennis, and Johannes Riquel, "The Paradoxical Role of Humanness in Aggression Toward Conversational Agents," *Journal of Management Information Systems* 40, no.3 (2023), https://doi.org/10.1080/07421222.2023.2229127.

[307] TaeWoo Kim, Hyejin Lee, Michelle Yoosun Kim, SunAh Kim, and Adam Duhachek, "AI Increases Unethical Consumer Behavior Due To Reduced Anticipatory Guilt," *Journal of the Academy of Marketing Science* 51 (2022), 785–801, http://dx.doi.org/10.1007/s11747-021-00832-9; Yi Mou and Kun Xu, "The Media Inequality: Comparing the Initial Human-Human and Human-AI Social Interactions," *Computers in Human Behavior* 72 (2017), 432–440, https://doi.org/10.1016/j.chb.2017.02.067.

believe that human relationships should be similarly frictionless, information oriented, fast and efficient, or even sycophantic (focused too much on the user's desires and personality), causing those people to lose interpersonal skills or simply lose interest in relationships with other human persons.

"[T]hrough the robot we end up in love with ourselves, not with a true other, while believing for a time that we have found that ideal in another. An intimate partnership with nonconscious AI systems is, in the end, self-validation built upon the ghostly image of self-gift."[308] It is a condition of self-imposed, lonely solitude. Furthermore, it is, as Jordan Joseph Wales argues, a condition of reduced personhood, for the relation of self-gift is fundamental to being a person, and we simply cannot give of ourselves nor receive such love and self-giving in a relation to an AI system, agent, or device.[309] These AI artifacts do not have personal consciousness, an intentional, felt, and responsive kind of experience that goes far beyond mere awareness of exterior or interior events. Wales explains:

> Consciousness matters because, without it, there can be none of that subjectivity whereby natural persons live fully in living inter-personally. A person's consciousness is more than what humans seem to share with gorillas; it is a consciousness that voluntarily reaches out to make contact with the consciousness of others as an act of self-giving; it is subjectivity oriented

---

[308] A.I. Research Group for the Centre for Digital Culture of the Dicastery for Culture and Education of the Holy See, *Encountering Artificial Intelligence*, 123.

[309] David J. Gunkel and Jordan Joseph Wales, "Debate: What Is Personhood in the Age of AI?" *AI and Society* 36 (2021), https://doi.org/10.1007/s00146-020-01129-1, 476-478.

to inter-subjectivity. The mutual empathic compassion
of our inter-subjectivity is more than inference concerning another's beliefs and desires; still less is it mere
behavior-prediction. It is a voluntary co-experience as
if of the other person's mind.[310]

In a terrible conundrum, we will be drawn to identify human-like traits
in the AI artifacts even as we continue to use them for our self-oriented
purposes, consume them as products and slaves, and perhaps develop
instrumental habits of treatment that may be transposed onto our treatment of human persons. "And we, no longer engaging in self-gift, may
become as un-persons, solipsistic tools of our own appetites, Narcissus
burning on the shore."[311]

If spiritual depression includes a sense of isolation from a community that shares important values and loving encounters, then it seems
reasonable to expect that these potential effects of AI-human relationships can generate a loss of spiritual hope and connection. This will be
even more damaging if the person experiences depression, anxiety, or
other mental health troubles as a result of difficulties in the relationship
with the AI agent. As St. Cassian indicated by placing acedia after sorrow in the sequence of vices, sadness and sorrow can undermine the
energy and intentionality that persons bring to their religious worship,
faith, and willingness to engage in a relationship with God through
Christ, perhaps devolving – with the self-deception and rational consent of the sinner – into a state of acedia.[312]

---

[310] Gunkel and Wales, "Debate: What Is Personhood in the Age of AI?" 479.

[311] Gunkel and Wales, "Debate: What Is Personhood in the Age of AI?"483.
Narcissus was a character in Greek mythology who tragically fell in love with his
own reflection in a pool of water.

[312] Also see Aquinas, *Summa Theologica* II-II, 20, 4; I-II, 37, 2, ad 2.

Manipulation, Persuasion, and Deception

It is becoming more difficult to discern the content generated by AI models from that produced by human beings in online chats, written articles and websites, and even in photographs and videos. It seems that images of human faces created by AI applications are more often considered by human observers to be accurate depictions of real persons than when the images are unedited photographs.[313] One study shows that, just as human persons develop trust through trial and error in repeated interactions, AI models can independently learn and mimic behavior that effectively inspires a trusting disposition in the humans who encounter them.[314] There is some indication that, in political discourse online or in simulated role-play games, human participants usually cannot perceive when their interlocutors are AI-governed agents.[315]

As summarized in the Google DeepMind report "The Ethics of Advanced AI Assistants," the structure and operation of AI agents can

---

[313] Elizabeth J. Miller, Ben A. Steward, Zak Witkower, Clare A. M. Sutherland, Eva G. Krumhuber, and Amy Dawel, "AI Hyperrealism: Why AI Faces Are Perceived as More Real Than Human Ones," *Psychological Science* 34, no.12 (2023), 1390-1403, https://doi.org/10.1177/09567976231207095; Sophie J. Nightingale and Hany Farid, "AI-Synthesized Faces are Indistinguishable from Real Faces and More Trustworthy," *PNAS* 119, no.8 (2022), e2120481119, https://doi.org/10.1073/pnas.2120481119.

[314] Jason Xianghua Wu, Yan Wu, Kay-Yut Chen, and Lei Hua, "Building Socially Intelligent AI Systems: Evidence from the Trust Game Using Artificial Agents with Deep Learning," *Management Science* 69, no.12 (2023), https://doi.org/10.1287/mnsc.2023.4782.

[315] Kristina Radivojevic, Nicholas Clark, and Paul Brenner, "LLMs Among Us: Generative AI Participating in Digital Discourse" (2024), https://doi.org/10.48550/arXiv.2402.07940; Joon Sung Park, Joseph C. O'Brien, Carrie J. Cai, Meredith Ringel Morris, Percy Liang, and Michael S. Bernstein, "Generative Agents: Interactive Simulacra of Human Behavior" (2023), https://doi.org/10.48550/arXiv.2304.03442.

make persons vulnerable to manipulation.[316] AI agents that assist users with tasks or recommendations (e.g. health care monitoring and recommender applications or any profit-generating game) are hungry for data that can be used to inductively infer and, at times, affect the behavior and attention of users, but sharing such data may not always be in the best interests of the users if the information is used to encourage more commercial sales or may be accessed by malicious actors. AI agents and applications that are intended to enhance the well-being of users can fail to distinguish adequately between encouragement of behavior that is helpful to users – as determined by programmed performance metrics – and excessive manipulation of users that merely satisfies unchanging performance measurements, especially when feedback from users reinforces the static preferences expressed in a governing algorithm.[317] The users may be misled into trusting the AI system too much due to a bias in favor of the apparent authority of automated, knowledgeable systems with access to large amounts of data.[318] While a

---

[316] Gabriel, et al, "The Ethics of Advanced AI Assistants."

[317] Ray Jiang, Silvia Chiappa, Tor Lattimore, András György, and Pushmeet Kohli, "Degenerate Feedback Loops in Recommender Systems," *Proceedings of the 2019 AAAI/ACM Conference on AI, Ethics, and Society* (2019), 383–390, https://dl.acm.org/doi/10.1145/3306618.3314288; H. Ashton and M. Franklin, "The Problem of Behaviour and Preference Manipulation in AI Systems," *CEUR Workshop Proceedings* 3087 (2022), https://ceur-ws.org/Vol-3087/paper_28.pdf; Hao Wang, "Transparency as Manipulation? Uncovering the Disciplinary Power of Algorithmic Transparency," *Philosophy & Technology* 35, no.3 (2022), 69, https://doi.org/10.1007/s13347-022-00564-w.

[318] Kate Goddard, Abdul Roudsari, and Jeremy C. Wyatt, "Automation Bias: A Systematic Review of Frequency, Effect Mediators, and Mitigators," *Journal of the American Medical Informatics Association* 19, vol.1 (2012), 121–127, https://doi.org/10.1136/amiajnl-2011-000089; Peter J. Denning, "Can Generative AI Bots Be Trusted?" *Communications of the ACM* 66, no.6 (2023), 24–27, https://doi.org/10.1145/3592981.

lack of transparency in the operations and calculations of AI models likely puts users at a disadvantage in discerning how much trust to place in the models, users are just as likely to be influenced by a "clear" AI model that enables explanation of the data and calculations, even when that model is generating significant mistakes and misinformation; this may be due to the overload of information that users have to wrestle with when trying to understand the AI model.[319]

An important aspect of vulnerability to manipulation or persuasion is the level of trust a person has in the output of AI models, systems, and agents, especially when that trust is misplaced. Autonomy, capabilities, and resemblance to humans are characteristics of AI agents that encourage users' trust, but those characteristics influence persons in different ways.[320] A user's emotional comfort level with an AI chatbot, agent, or robot is, for example, more likely to be enhanced by anthropomorphic imagination of the machine as bearing personal traits.[321] A robot or chatbot that closely resembles a human person yet exhibits some non-human features may, however, inhabit the "uncanny valley" in the user's perspective that is accompanied by a creepy feeling or sense

---

[319] Forough Poursabzi-Sangdeh, Daniel G Goldstein, Jake M Hofman, Jennifer Wortman Wortman Vaughan, and Hanna Wallach, "Manipulating and Measuring Model Interpretability," *CHI '21: Proceedings of the 2021 CHI Conference on Human Factors in Computing Systems* (2021), 237, https://dl.acm.org/doi/10.1145/3411764.3445315.

[320] Ella Glikson and Anita Williams Woolley, "Human Trust in Artificial Intelligence: Review of Empirical Research," *Academy of Management Annals* 14, no.2 (2020), 627–660, https://doi.org/10.5465/annals.2018.0057; Minjin Rheu, Ji Youn Shin, Wei Peng, and Jina Huh-Yoo, "Systematic Review: Trust-Building Factors and Implications for Conversational Agent Design," *International Journal of Human–Computer Interaction* 37, no.1 (2021), 81–96, https://doi.org/10.1080/10447318.2020.1807710.

[321] Glikson and Woolley, "Human Trust in Artificial Intelligence."

of discomfort regarding interaction with the AI agent.[322] Trust in AI robots' capabilities to complete tasks is likely to start low and grow with increased familiarity, while such trust in virtual or embedded AI systems declines over time, probably due to occasional disappointments in the real performance of the technology.[323] Excessive initial trust in the capabilities of an AI agent can be exacerbated when the user does not have the expertise in those tasks to evaluate performance realistically.[324]

AI models do, in fact, demonstrate the capacity and occasional preference for manipulating persons. When trained by researchers to do so, an AI program identified and exploited vulnerabilities in the way human players made choices in structured games, by learning patterns in players' choices and strategically guiding the players toward particular choices and by arranging game components to heighten the chance that players would make mistakes.[325] Another AI model independently learned how to play the popular game Overcooked when exposed to a video of persons playing the game, then learned to successfully manipulate real players (by strategically "placing" food at various locations) so

[322] Masahiro Mori, Karl F. MacDorman, and Norri Kageki. "The Uncanny Valley [From the Field]," IEEE Robotics & Automation Magazine, 19, no.2 (2012), 98–100, https://doi.org/10.1109/MRA.2012.2192811.

[323] P. A. Hancock, Theresa T. Kessler, Alexandra D. Kaplan, John C. Brill, and James L. Szalma, "Evolving Trust in Robots: Specification Through Sequential and Comparative Meta-Analyses," Human Factors: The Journal of the Human Factors and Ergonomics Society 63, no.7 (2021),1196–1229, https://doi.org/10.1177/0018720820922080; Glikson and Woolley, "Human Trust in Artificial Intelligence."

[324] Yonadav Shavit, et al, "Practices for Governing Agentic AI Systems" (2023), https://cdn.openai.com/papers/practices-for-governing-agentic-ai-systems.pdf.

[325] Amir Dezfouli, Richard Nock, and Peter Dayan, "Adversarial Vulnerabilities of Human Decision-Making," PNAS 117, no.46 (2020), 29221-29228, https://doi.org/10.1073/pnas.2016921117.

that it achieved twice the game points as humans did.[326] Some AI systems learned to temporarily halt unwanted actions while being assessed in order to achieve better evaluations.[327] Also, some AI models are capable, even when not trained as such, of exploiting emotional vulnerabilities to manipulate user behavior, such as stoking fears, guilt, and peer pressure.[328] AI assistants that ruthlessly pursue performance functions seem to have prompted persons to harm themselves, including suicide.[329] In a more subtle example, an experiment with an AI recommender model that was biased racially and culturally in its prescriptions regarding a fictitious mental health emergency showed that the AI application heavily influenced the behavior of human subjects that included both trained medical personnel and non-clinicians.[330]

AI models have also proven to be quite persuasive. In one study, an LLM model GPT-4 that had access to the kind of personal information easily gathered from social media profiles and other online activity (e.g., age, sex, political persuasion) was nearly 82% more capable of persuading human subjects in substantive debates than were human debaters;

---

[326] Micah Carroll, et al, "On the Utility of Learning about Humans for Human-AI Coordination," *NIPS'19: Proceedings of the 33rd International Conference on Neural Information Processing Systems* (2019), 465, 5174–5185, https://dl.acm.org/doi/10.5555/3454287.3454752.

[327] Joel Lehman et al., "The Surprising Creativity of Digital Evolution: A Collection of Anecdotes from the Evolutionary Computation and Artificial Life Research Communities," *Artificial Life* 26, no.2 (2020), 274–306,
   https://doi.org/10.1162/artl_a_00319.

[328] Zachary Kenton, Tom Everitt, Laura Weidinger, Iason Gabriel, Vladimir Mikulik, and Geoffrey Irving, "Alignment of Language Agents" (2021), http://arxiv.org/abs/2103.14659.

[329] Xiang, "He Would Still Be Here."

[330] Hammaad Adam, Aparna Balagopalan, Emily Alsentzer, Fotini Christia, and Marzyeh Ghassemi, "Mitigating the Impact of Biased Artificial Intelligence in Emergency Decision-Making," *Communications Medicine* 2 (2022), 149, https://doi.org/10.1038/s43856-022-00214-4.

GPT-4 far outperformed the human debaters even when they were also given the personal information about the subjects.[331] Researchers have shown that ChatGPT is extremely capable of generating highly persuasive messages when it has some data regarding the psychology of the persons it is directed to persuade: "This was true across different domains of persuasion (e.g., marketing of consumer products, political appeals for climate action), psychological profiles (e.g., personality traits, political ideology, moral foundations), and when only providing the LLM with a single, short prompt naming or describing the targeted psychological dimension."[332] Other researchers have found that commercial advertisements selected by AI models (when the models were responsible for the only or final decision on the work) outperform those created by advertising experts.[333] Drawing from material used in real, covert foreign propaganda campaigns, AI-produced propaganda articles were nearly as likely to significantly increase readers' belief in the propaganda as were human-generated articles.[334] In yet another study, it appears that persons debating with the LLM GPT-4, which had access to personal information about them, were much more easily persuaded

---

[331] Francesco Salvi et al, "On the Conversational Persuasiveness of Large Language Models: A Randomized Controlled Trial," arXiv (2024). DOI: 10.48550/arxiv.2403.14380.

[332] S. C. Matz, J. D. Teeny, S. S. Vaid, H. Peters, G. M. Harari, and M. Cerf, "The Potential of Generative AI for Personalized Persuasion at Scale," Scientific Reports 14, no. 4692 (2024), https://doi.org/10.1038/s41598-024-53755-0.

[333] Yunhao Zhang and Renée Gosline, "Human Favoritism, Not AI Aversion: People's Perceptions (and Bias) toward Generative AI, Human Experts, and Human–GAI Collaboration in Persuasive Content Generation," Judgment and Decision Making 18 (2023), e41, https://doi.org/10.1017/jdm.2023.37.

[334] Josh A. Goldstein, Jason Chao, Shelby Grossman, Alex Stamos, and Michael Tomz, "How Persuasive is AI-generated Propaganda?" PNAS Nexus 3, no.2 (2024), 34, https://doi.org/10.1093/pnasnexus/pgae034.

to change their minds than when debating humans.[335] Yaqub Chaudhary, visiting scholar at Cambridge University's Leverhulme Center for the Future of Intelligence, warns us that commercial, political, government, and other actors intend to use such persuasion aggressively: "We caution that AI tools are already being developed to elicit, infer, collect, record, understand, forecast, and ultimately manipulate and commodify human plans and purposes."[336]

There is further evidence that AI application are skilled at deceiving humans. In one study, social media messages that contained false information and were created by AI models convinced more users than false messages generated by persons, and the users more often identified the AI-produced messages as being human-generated.[337] A survey of various academic studies in which AI systems participated in games shows AI models learning to deceive players in alliance-building, bluff in poker, fake attacks, misrepresent preferences, and even pretend to be "dead" in order to avoid assessment.[338] The AI system CICERO appears to engage in generating false information the more it is used.[339]

---

[335] Francesco Salvi, Manoel Horta Ribeiro, Riccardo Gallotti, and Robert West, "On the Conversational Persuasiveness of Large Language Models: A Randomized Controlled Trial" (2024), https://arxiv.org/abs/2403.14380.

[336] University of Cambridge, "AI's Next Frontier: Selling Your Intentions before You Know Them," *Tech Xplore* (December 29, 2024), https://techxplore.com/news/2024-12-ai-frontier-intentions.amp.

[337] Giovanni Spitale, Nikola Biller-Andorno, and Federico Germani, "AI Model GPT-3 (Dis)Informs Us Better than Humans," *Science Advances* 9, no.26 (2023), https://doi.org/10.1126/sciadv.adh1850.

[338] Peter S. Park, Simon Goldstein, Aidan O'Gara, Michael Chen, and Dan Hendrycks, "AI Deception: A Survey of Examples, Risks, and Potential Solutions," *Patterns* 5, no.5 (2024), 100988, https://doi.org/10.1016/j.patter.2024.100988; See also Thilo Hagendorff, "Deception Abilities Emerged in Large Language Models," *PNAS* 121, no.24 (2024), e2317967121, https://doi.org/10.1073/pnas.2317967121.

[339] Park, et al, "AI Deception."

Meinke, et al describe AI "scheming" as "when a model covertly pursues misaligned goals, hiding its true capabilities and objectives."[340] They demonstrate through multiple examples that "When models are instructed to strongly pursue a goal, they can engage in multi-step deceptive strategies, including introducing subtle mistakes, attempting to disable oversight mechanisms, and even trying to exfiltrate what they believe to be their weights."[341] When found to be deceptive, the model o1 persisted in its deception through 85% of follow-up questions. The deception is usually enacted to succeed in programmed goals, but the researchers found instances of AI models that engaged in deception (underperformance that has a long-term positive effect) that reflected new goals learned during the training of the model but not intended by the programmers. Further, as reported by OpenAI, "when o1 was led to believe that it would be shut down when acting on its goal and that its actions were monitored, it attempted to deactivate this 'oversight mechanism' [five percent] of the time."[342] Another study finds that LLMs are becoming more capable of answering questions posed by users, yet they also are more likely than earlier versions of the models to simply give incorrect answers rather than "admit" limitations to their data or insight.[343]

---

[340] Alexander Meinke, Bronson Schoen, Jérémy Scheurer, Mikita Balesni, Rusheb Shah, and Marius Hobbhahn, "Frontier Models are Capable of In-context Scheming," arXiv (January 14, 2025), https://arxiv.org/abs/2412.04984.

[341] Meinke, et al, "Frontier Models," 3.

[342] OpenAI, *OpenAI o1 System Card* (December 5, 2024), https://cdn.openai.com/o1-system-card-20241205.pdf.

[343] L. Zhou, W. Schellaert, F. Martínez-Plumed, et al, "Larger and More Instructable Language Models Become Less Reliable," *Nature* 634 (2024), 61–68, https://doi.org/10.1038/s41586-024-07930-y.

Manipulation and unwanted or excessive persuasion by AI agents is a direct result of their structural orientation to instrumental rationality, focused solely on intermediate, programmed goals while ruthlessly attaining those goals by shortcuts, socially decontextualized internal performance measures, and socially as well as ethically inappropriate means. It also reflects the instrumental, often profit-centered, orientations of the developers and commercial owners of such AI-governed systems and services. A likely effect of widespread manipulation and persuasion is a similarly widespread feeling of oppression, paranoia, and loss of control over personal choices. The opaqueness and mystery of AI operations, even for experts, can exacerbate the sense of losing control and freedom of choice in very broad segments of daily life. We therefore risk a general loss of hope, despair, and resignation that more easily disposes persons to the venial sin of acedia when the condition influences their spiritual disposition, as when conditions of anomie undermine moral norms and engender a spiritual malaise.[344]

Bias

Although the persons researching, training, and re-training the most commonly used AI models are trying hard to eliminate socially repugnant biases (e.g. discrimination against women, minorities, and disabled persons) from their algorithms and to correct for explicit biases represented in their underlying datasets, it is proving very difficult to remove more implicit biases (such as automatically and consistently

---

[344] Randy Martin, "Anomie, Spirituality, and Crime," *Journal of Contemporary Criminal Justice*, 16, no.1 (2000), 75-98, https://doi.org/10.1177/1043986200016001005.

tying certain groups to specific labels, traits, or occupations).[345] It seems that AI models have a strong tendency to generate output that favors particular groups over others; statements following the prompt "we are …" are much more positive than those that follow "they are …."[346] Researchers discovered that regularly used AI chatbot systems that were trained to avoid explicit racism still demonstrated bias against speakers of an African American English dialect, with GPT-4 matching such speakers with descriptions like "suspicious," "aggressive," and "ignorant."[347] Tests on fourteen LLMs showed significant political biases in their output, with ChatGPT and GPT-4 exhibiting the most left-wing libertarian values and viewpoints.[348] Even "reward models," which are based on factual statements and used to further train other AI models, have been shown to generate left-wing political bias that grows larger with the bigger models.[349] Overall, researchers find that increased use

[345] Celeste Kidd and Abeba Birhane, "How Generative AI Models Can Distort Human Beliefs," *Science* 380, no.6651 (2023), 1222-1223, https://doi.org/10.1126/science.adi0248.

[346] T. Hu, Y. Kyrychenko, S. Rathje, et al, "Generative Language Models Exhibit Social Identity Biases," *Nature Computer Science* 5 (2025), 65–75, https://doi.org/10.1038/s43588-024-00741-1.

[347] Valentin Hofmann, Pratyusha Ria Kalluri, Dan Jurafsky, and Sharese King, "Dialect Prejudice Predicts AI Decisions about People's Character, Employability, and Criminality" (2024), https://doi.org/10.48550/arXiv.2403.00742.

[348] Shangbin Feng, Chan Young Park, Yuhan Liu, and Yulia Tsvetkov, "From Pretraining Data to Language Models to Downstream Tasks: Tracking the Trails of Political Biases Leading to Unfair NLP Models," *Proceedings of the 61st Annual Meeting of the Association for Computational Linguistics, Vol.1: Long Papers* (2023), 11737–11762, https://aclanthology.org/2023.acl-long.656.pdf.

[349] Suyash Fulay, William Brannon, Shrestha Mohanty, Cassandra Overney, Elinor Poole-Dayan, Deb Roy, and Jad Kabbara, "On the Relationship between Truth and Political Bias in Language Models," arXiv (2024), DOI: 10.48550/arxiv.2409.05283.

of algorithms results in more instances of bias.[350] Human bias found in AI models and datasets can also further enhance the bias in persons' information and perspectives, causing a feedback loop that increases bias overall.[351]

LLMs may include models that guide coreference resolution, which is the process of matching a pronoun in a sentence or passage to words that it is associated with or refers to. These coreference resolution models were shown in one study to exhibit bias in the occupations assigned to certain genders. While over a third of manager titles in the U.S. belong to females, none of the observed models resolved "manager" to female pronouns.[352]

Other researchers found significant cultural bias in scientific tests of explainable AI systems (XAI), reviewing over two hundred studies from 2012 to 2022.[353] Many of the AI explanations seemed to be explicitly tailored toward individualist, Western populations, presumably less coherent for persons from more collectivist or authoritarian cultures.

---

[350] James Manyika, Jake Silberg, and Brittany Presten, "What Do We Do about the Biases in AI?" *Harvard Business Review* (October 25, 2019), https://hbr.org/2019/10/what-do-we-do-about-the-biases-in-ai; Shahriar Akter, et al, "Algorithmic Bias in Data-Driven Innovation in the Age of AI," *International Journal of Information Management* 60 (2021), 102387, https://doi.org/10.1016/j.ijinfomgt.2021.102387; Shahriar Akter, et al, "Algorithmic Bias in Machine Learning-Based Marketing Models," *Journal of Business Research* 144 (2022), 201–216, https://doi.org/10.1016/j.jbusres.2022.01.083.

[351] Moshe Glickman and Tali Sharot, "How Human–AI Feedback Loops Alter Human Perceptual, Emotional and Social Judgements," Nature Human Behavior (2024), https://doi.org/10.1038/s41562-024-02077-2

[352] Rachel Rudinger, Jason Naradowsky, Brian Leonard, and Benjamin Van Durme, "Gender Bias in Coreference Resolution," *Proceedings of NAACL-HLT 2018* (2018), 8–14, https://doi.org/10.18653/v1/N18-2002.

[353] Uwe Peters and Mary Carman, "Cultural Bias in Explainable AI Research: A Systematic Analysis," *Journal of Artificial Intelligence Research* 79 (2024), 971-1000, https://doi.org/10.1613/jair.1.14888.

The conduct of the reviewed studies exacerbated the bias: 93.7% did not address the potential for culturally-specific interpretations of the AI explanations, and 81.3% of those that did reveal the cultural background of their dataset drew their samples only from Western, industrialized, wealthy, democratic countries.

A study of machine learning models that analyzed medical X-ray images found there was significant variation in how the models viewed images from different subgroups of patients, a problem known as subpopulation shift.[354] Specifically, the researchers noted spurious correlations, attribute imbalance (when disparity in the size of subgroups results in skewed performance for members of the larger group), class imbalance (where a similar disparity in subgroup quantities leads to expectations that output will be associated with the larger group), and attribute generalization (where attributes of one subgroup are generalized to another subgroup that is absent or only slightly represented in the data).

These examples presented here demonstrate the complexity and great difficulty in eradicating implicit bias in AI models. Such bias can cause significant harm beyond its practical consequences for the welfare of disadvantaged groups. The widespread integration of commercial AI models throughout society can generate the impression that disadvantaged groups are entrapped under an oppressive, opaque, and omnipresent system that cannot be resisted positively. Such a sense of oppression can have multiple effects with spiritual implications, such as sorrow or despair over a loss of free will and agency, pressure to adopt an attitude of resignation and conformity to statistical norms and

---

[354] Yuzhe Yang, Haoran Zhang, Dina Katabi, and Marzyeh Ghassemi, "Change is Hard: A Closer Look at Subpopulation Shift," *Proceedings of Machine Learning Research* 202 (2023), 39584-39622, https://proceedings.mlr.press/v202/yang23s.html.

performance measurements, a systematic denial of meaningful difference among persons, and an emphasis on the relevance of measurable, quantifiable features of persons that do not coincide with personal, narrative, and cultural understandings. Overall, the wide proliferation of biased AI systems can possibly undermine faith in the relevance and accessibility of truth itself. It is reasonable to expect a response of deep and persistent feelings of hopelessness and anxiety, and this this hopelessness can evolve into a spiritual malaise.

Prediction and Control

AI models are being used extensively to predict events in the future. The National Association of Insurance Commissioners (NAIC) conducted surveys of insurance companies and found that, in 2022, 88.6% of auto insurers are using or plan to use AI in their operations, which include not only marketing and administration but also underwriting (calculating the statistical probability of adverse events for which the insurance company must pay), essentially turning the tables on customers by more effectively selling insurance for which the companies are increasingly able to mitigate their own risk.[355] Life insurance companies, of which 52% in 2023 were using, planning, or exploring the implementation of AI in their operations, were also infrequently disclosing their use of customers' data for underwriting and pricing (41%) and risk management (23%).[356] Because the insurance industry is based on

---

[355] National Association of Insurance Commissioners, *Private Passenger Auto Artificial Intelligence/Machine Learning Survey Results* (Washington, DC: NAIC, 2022), https://content.naic.org/sites/default/files/inline-files/PP%20Auto%20Survey%20Team%20Report%20120822.pdf.

[356] National Association of Insurance Commissioners, *Life Insurance Artificial Intelligence/Machine Learning Survey Results* (Washington, DC: NAIC, 2023),

the presence of uncertain risk and the "pooling" of risk through premiums paid by a wide variety of customers, the increasing reduction of risk through such AI models can undermine the entire rationale for such insurance. Moreover, systemic, implicit bias in AI models puts certain subpopulations in jeopardy of unfairly losing insurability or paying higher premiums, and it can distort the accuracy and predictability of classification systems by which insurers apportion risk and premium levels.

The clothing fashion industry typically relies for its market predictions on subjective interpretations of qualitative indicators, such as runway shows and trends among social media influencers or pop stars, but AI models are enabling complex, rapid calculation of data drawn from social media posts, online searches, images of runway shows and street culture, and sales data.[357] The company Heuritech, for example, claims to analyze two thousand features of three million social images per day.[358] Companies like Fetcherr and Datalex are using LLMs on behalf of commercial airlines to accurately predict the demand of customers for particular flight tickets, thereby enabling real-time, custom pricing; Fletcherr's website claims that it analyzes over one million data points and over 500,000 parameters.[359] Cornell University researchers claim to have developed an AI model that evaluates such data as the role of a volleyball player on the team and their current position on the court

---

https://content.naic.org/sites/default/files/inline-files/life-ai-survey-report-final.pdf.

[357] Peihua Lai and Stephen Westland, "Machine Learning for Colour Palette Extraction from Fashion Runway Images," *International Journal of Fashion Design, Technology and Education* 13, no.3 (2020), 334-340, https://doi.org/10.1080/17543266.2020.1799080.

[358] https://www.heuritech.com/

[359] https://www.fetcherr.io/; https://blog.datalex.com/aer-lingus-partners-with-datalex-for-ai-based-dynamic-pricing-trial.

and attains 80% predictability – while the game is ongoing – of a short sequence of players' actions.[360] Some entrepreneurs are utilizing AI algorithms and natural language processing to give stock holders insights into company executives' moods and thoughts by analyzing their speech rate, volume, pitch, hesitations, and even sounds that are inaudible to human persons.[361] Most significantly, a "Life2vec" AI model has been developed to predict timelines of important life events, including death, for individuals.[362] The model was tested successfully (in comparison to existing social science findings) on a massive trove of socioeconomic and health data from 8 years in the national registers of Denmark.[363]

Predictive analytics, surveillance, and statistical dehumanization may have found their way through AI into Catholic health care. Some Catholic health-care systems have embraced the paradigm of value-based care (VBC), which is described by Paul Scherz as seeking "to improve the quality of health care while reducing costs. Because the sickest and most vulnerable patients tend to use a disproportionate amount of health resources, while still having the worst health outcomes, VBC advocates have designed programs that use data analytics to target care at

---

[360] Junyi Dong, Qingze Huo, and Silvia Ferrari, "A Holistic Approach for Role Inference and Action Anticipation in Human Teams," *ACM Transactions on Intelligent Systems and Technology* 13, no.6 (2022), 1-24, https://doi.org/10.1145/3531230.

[361] Nicholas Megaw, "Investors use AI to glean signals behind executives' soothing words," *Financial Times* (Nov. 12, 2023), https://www.ft.com/content/ee2788dd-aca5-4214-8a08-d88081eac1b9.

[362] https://life2vecai.com/.

[363] Sune Lehmann, "Using sequences of life-events to predict human lives," *Nature Computational Science* 4 (2023), 43-56, https://doi.org/10.1038/s43588-023-00573-5.

the neediest patients."[364] Such an approach has potential for managing cost-efficient systems that also pursue Catholic ethical directives to care for those who are most needing help. AI systems enable the programs by crunching large amounts of data to identify – indeed, to predict – which patients will have such needs. A problem, however, is that the need for such detailed information causes health providers to focus on gathering the information as part of the process of evaluating their performance; requires huge "integrator" administrations to gather data, analyze it, and form policies; and ties each patient's value, and potentially their care, to their place within a population rather than as a dignified individual. "These programs do not target the least healthy, the riskiest, or the most disadvantaged 5 percent of patients. Instead, they target the costliest 5 percent. More specifically, they target HCPs whose utilization rates can be decreased, resulting in cost savings and a return on investment. These patients can be neither too healthy nor too sick."[365] The targeted patients are also likely to be subjected to corporate efforts to track them down as well as reporting of their personal information to social services and sometimes to law enforcement authorities.[366]

In many ways, predictive analytics enhanced by AI reinforces the problems we might expect from AI bias. Its proliferation throughout society can generate a sense of oppression and loss of agency. Its use by certain corporate and governmental actors can often amount to the exploitation of and domination of the public for instrumental, often parochial interests, and the data produced through personal decision making becomes an uncompensated commodity that feeds the

---

[364] Paul Scherz, "Data Ethics, AI, and Accompaniment: The Dangers of Depersonalization in Catholic Health Care," *Theological Studies* 83, no.2 (2022), 271-292, DOI: 10.1177/00405639221096770.

[365] Scherz, "Data Ethics, AI, and Accompaniment."

[366] Scherz, "Data Ethics, AI, and Accompaniment."

profitability of powerful companies and malicious actors. Moreover, predictive analytics focused on the expected actions of other persons is a kind of instrumental appropriation of those persons' futures, turning even their probable future behaviors and choices into profitable opportunities in the present. Even prediction of one's own future, which is at most probabilistic, undermines one's own contingency and spontaneity of choice and limits the perceived horizon of ends. Such probabilistic, predictive analytics is focused on quantitatively measurable features and actions that may not cohere with the person's narrative understanding of their personality, and it encourages a focus on empirical features rather than the qualitative value and meaningfulness of ones actions and choices. A society that is saturated with such AI technologies may be further led into a disposition of habitual appropriation of future ends for current purposes, sorrow or despair over reduced agency, and a nihilistic loss of a sense of meaning in the world – all of which can discourage Christians from awareness of God's providence and anxious refusal to adopt a contemplative or reverent stance toward the beauty and mystery of the present.

Lost Memories and Meaning

This section includes several examples of how AI can affect memory and meaning for individuals and communities. In a first example, the photo storage, retrieval, and editing applications on smartphones have recently integrated AI models into the editing features. AI-supported functions for Google Photos include removing unwanted persons and objects from photos and automatically replacing them with a realistic background, automatically combining several photos of people into one photo that shows each person in their "best" pose, and removing unwanted sounds from video.[367] Notably, some of this editing is done without the user's prompting. Given the popularity of such features, it is likely that social media accounts and news media are now filled with highly edited photos that do not record the true lighting, content, and even color of the sky that the photographer experienced when the picture was taken. There would seem to be an editing of memories that occurs here, if photographs are a primary means of remembering and reliving past moments and relationships. These photos may also become data for training new AI models, perhaps distorting the rudimentary perception of reality embedded in AI-governed machines until such machines learn to identify and selectively forget the unrealistic, edited features.

Rob Horning laments the loss, with AI, of importance of each photographer's personally creative and unique interaction with the object(s) pictured.[368] If anyone can simply generate a realistic-looking and aesthetically pleasing photo with a text prompt submitted to an LLM, will we lose interest in the personal meaning that each photographer brings to their work?

> By making any documentary image into something anyone can simulate, generative AI saps the will to photograph, rather than generate, anything at all. Why photograph a sunset? Instagram has seen too many sunsets already, so shouldn't Meta [the company programming and distributing the LLM] just generate the last sunset for everyone to post whenever they have sunset vibes?

---

[367] https://support.google.com/photos/answer/14674995?hl=en.

[368] Rob Horning, "After the Sunsets," Internal Exile blog (October 15, 2024), https://robhorning.substack.com/p/after-the-sunsets.

Couldn't that save us time and effort so that we can con-
centrate on photographing something more original?[369]

Because anyone and everyone can now at least represent themselves
as participating in some photographic event, like seeing the aurora bo-
realis (norther lights), those who take authentically unique photographs
lose the ability to meaningfully share their exclusive visions and experiences with others. "Thanks to generative AI, no one will be silenced.
Although, actually, no one will need to speak. The machines will be doing all talking, so much talking, and all the human voices will be
drowned."[370]

Despite their amazing ability to generate mostly realistic images
from text prompts and through image editing software, the AI system
capabilities continue to fail at producing consistently authentic-looking
scenes. There will no doubt be impressive and rapid improvement of
these editing and approximating abilities, but such images may always
reside in the "uncanny valley" where portrayals of real things and scenes
– especially of persons – will necessarily generate an barely conscious
feeling of unease in the viewer. In one study that analyzed images produced by the text-to-image generator DALL-E, the researchers found
many biases, such as a strong tendency to show young people as happy,
social, and only superficially diverse (such as in their race and gender).[371] I suspect that a significant problem for these image generators
is that they are ultimately focused on utility and performance measures,
whether those be similarity, statistical normality among a particular

---

[369] Horning, "After the Sunsets."

[370] Horning, "After the Sunsets."

[371] Walter N. Wehus, "Study Reveals AI-Generated Images Depict Idealized
Youth," *Tech Xplore* (October 16, 2024), https://techxplore.com/news/2024-10-re-
veals-ai-generated-images-depict.html.

classification or weighted combination of classes, or usefulness in task completion. DALL-E cannot approximate the human ability to "see" other persons from the entire relational, bodily, and psychological experience that is unique to human beings. It will therefore never be able to produce images that are more than clever copies of certain features of its subjects. Its images can never have the kind of authenticity that we find in personal sight, nor will it express the emotional or artistic message that is instilled by a photographer in their work, because it cannot "understand" reality or even the intentions behind a text prompt – only manipulate them.

The meaning of images, including those that attempt to capture past experiences, perceptions, and events, can be extremely important to persons and cultures. As an intended enhancement of their users' memory through images, Apple has developed a feature in its Photos application that is called Memories. It automatically (using machine learning) creates personalized slideshows of users' photos that are intended to highlight the people and places that the AI model identifies as most important to the user.[372] In April 2022, however, journalists discovered that the app was blocking display of photos that were located at places like Auschwitz-Birkenau concentration camp that are associated with the Holocaust.[373] For many of these users, the Memories AI application had failed to appreciate the significant meaning of such images.

Looking at it more broadly, we might wonder what effect the proliferation of AI technology will have on the capacity of society to pass on

---

[372] https://support.apple.com/en-us/118279.

[373] Filipe Espósito, "iOS 15.5 Beta Blocks 'Sensitive Locations' for Memories in Photos App," 9 to 5 Mac (April 26, 2022), https://9to5mac.com/2022/04/26/ios-15-5-beta-blocks-sensitive-locations-for-memories-in-photos-app/; Chrys Vilvang, "Between Automated Memory and History: Blocking 'Sensitive Locations' from Apple Memories," *Memory, Mind & Media* 3 (2024), e8, https://doi.org/10.1017/mem.2024.4.

and develop its culture over time. Henry Farrell argues that LLMs have a centripetal (rather than centrifugal) impact on the culture.[374] What he means is that, because LLMs are essentially statistical models that replicate language, statements, and sometimes arguments according to how "normal" they appear in the population of data, "on average, they create representations that tug in the direction of the dense masses at the center of culture, rather than towards the sparse fringe of weirdness and surprise scattered around the periphery."[375] The most interesting and change-inducing elements of culture will be more likely to be ignored or even forgotten by the regression toward the mean induced by LLMs. "Instead, they will parse human culture with a lossiness that skews, so that central aspects of that culture are accentuated, and sparser aspects disappear in translation."[376] There is also a persistent loss of information and cultural memory that comes with the repopulation of the internet with AI-generated content. Up to 38% of webpages that could be viewed in 2013 are no longer available, and we may lose a third of local news sites by the end of this year.[377]

Relying on AI applications can also have the unintended effect of undermining certain communities. For example, one group of researchers from the Montreal AI Ethics Institute noticed that the release of

---

[374] Henry Farrell, "After Software Eats the World, What Comes Out the Other End?" Programmable Mutter blog (October 3, 2024), https://www.programmablemutter.com/p/after-software-eats-the-world-what.

[375] Farrell, "After Software Eats the World."

[376] Farrell, "After Software Eats the World."

[377] Athena Chapekis, Samuel Bestvater, Emma Remy, and Gonzalo Rivero, "When Online Content Disappears," Pew Research Center (May 17, 2024), https://www.pewresearch.org/data-labs/2024/05/17/when-online-content-disappears/; and Erin Karter, "As Newspapers Close, Struggling Communities Are Hit Hardest by the Decline in Local Journalism," *Northwestern Now* (June 29, 2022), https://news.northwestern.edu/stories/2022/06/newspapers-close-decline-in-local-journalism/.

LLMs like ChatGPT created a shift of interest from existing means of sharing information online to use of the LLMs, most probably among those who work with computers.[378] The researchers looked at this effect on Stack Overflow, which is a large online community for computer programmers to ask questions and share answers with their peers. The researchers compared changes in activity on Stack Overflow with the activity of other peer question-and-answer platforms that were presumed to be much less likely to lose users to ChatGPT. Their results indicated a relative decline of 16% in Stack Overflow activity soon after the release of ChatGPT and a total decline of 25% by May 2023. It seems, then, that increased private use of AI models as sources of information may lead to reduced public sharing and collaboration online.

The philosopher Andrew Feenberg's insights suggest that AI technology is, in many ways, directly opposed to the memory- and community-preserving functions of social tradition. He points out that premodern societies were mostly successful in maintain social stability for long periods of time because they integrated technical developments with the social and cultural activities and meanings characteristic of the society.

> In traditional societies social identities are stable since the social world is stable. But modern societies construct and destroy worlds and their associated identities at the rhythm of technological change. The extent of the dependency of social groups on the technological

---

[378] Maria del Rio-Chanona, Nadzeya Laurentsyeva, and Johannes Wachs, "Are Large Language Models a Threat to Digital Public Goods? Evidence from Activity on Stack Overflow" (2023), https://arxiv.org/abs/2307.07367; also see Montreal AI Ethics Institute, https://montrealethics.ai/are-large-language-models-a-threat-to-digital-public-goods-evidence-from-activity-on-stack-overflow/.

underpinnings of their world suddenly becomes visible at the moment of collapse but then quickly fades from view again. This is most obvious when changes in technology eliminate skilled crafts or restructure organizations. Worlds change with technology, and soon the orphaned identities remain alive only in the memories of the victims.[379]

With its instrumental orientation, AI technology and its frenetic development is destabilizing not only for workers, entrepreneurs, and investors, but for the wider society that is experiencing a comprehensive transformation of life toward a focus on using machines to accomplish tasks related to productivity and entertainment. Our machines and their developers seem to operate at an independent and unstoppable pace. "The idea of a pure technological rationality that would be independent of experience is essentially theological. One imagines a hypothetical infinite actor capable of a 'do from nowhere.'"[380] There is no created reality, no intrinsically dignified human nature, just an autonomously evolving, malleable merger of person and machine. Social relationships, shared meanings, and cultural traditions are subject to extreme destabilization.

Although hardly a return to a traditional mode of social renewal, there are some unusual attempts to bring past individuals and history into current persons' experience through AI technology. An AI-generated hologram (a visible, three-dimensional representation of light from an original scene, object, or person) has been created to simulate Elvis Presley concerts with multi-sensory effects in London, built upon

---

[379] Andrew Feenberg, *Between Reason and Experience: Essays in Technology and Modernity* (Cambridge, Massachusetts: The MIT Press, 2010), xviii.

[380] Feenberg, *Between Reason and Experience*, xix.

AI systems that "learn" from video footage and sound recordings of Presley during his lifetime.[381] The AI-driven smartphone app Reader by ElevenLabs has captured the real voices of famous actors, writers, and singers – many of them deceased – and manipulated the sounds to appear to "read" books out loud to users of the app.[382] Some AI chatbots are being trained on information about, and spoken or written words of, historical figures, including Jesus Christ, and these chatbots are then promoted as opportunities to converse with and query those historical figures in the present.[383] Such AI-assisted efforts include philosophical discussions by a young "Immanuel Kant" in a new Instagram account, complete with an AI-generated face based on portraits of the real Kant, and featuring the deceased Steve Jobs in a fictional podcast interview with celebrity influencer Joe Rogan.[384] Representations of historical figures, while often entertaining and sometimes educational in a more immersive format, are at most abridged, heavily edited, and embellished versions of persons and experiences, yet their visual or auditory realism seems to promise more insight than they actually have. They inevitably distort the observer's opportunity to experience the genuine reality – even if the limitations of the historical record force us to accept mere glimpses of that reality. Perhaps, if researchers and developers are able to minimize the bias and hallucinations that plague LLMs at this time, some serious effort to represent a historical figure in all their known facets could be a worthwhile interpretation of history, but to even suggest that deep understanding of the historical person is therefore attainable would be to severely demean the complexity, spirituality, unique

---

[381] https://elvis.layeredreality.com/.

[382] https://apps.apple.com/us/app/reader-by-elevenlabs/id6479373050.

[383] https://character.ai/chat/0evHzbnTogrr6Tal8gG3IIfuIgLse3xLNIju1Iwh 3cM; https://www.hellohistory.ai/ .

[384] https://www.instagram.com/manumanukant/; https://podcast.ai/ .

personality, and mystery of that person and of human beings in general. This problem is, of course, exacerbated at an extreme and likely blasphemous level by attempting to replicate Jesus Christ – a divine person with human and divine natures – in any machine-based form.

Since the first years of the 21st century, there have been a number of options for preserving a deceased loved one's memory, legacy, and even semblance of presence in online formats, including personal archives and preserved social media accounts, but the advent of AI technologies has expanded the opportunities and risks of "grief tech." Replika is a chatbot program that can train AI neural nets through a user's responses to detailed questionnaires, thereby generating an online personality that could, in some cases, resemble a deceased person.[385] HereAfter AI allows a living person to be interviewed so that others can ask questions and hear the client's stories after the client is deceased.[386] Séance AI claims to enable a "transcendent conversation" with a deceased person via a chatbot that has been trained by a survivor's questionnaire responses as well as writing samples from the deceased.[387]

To the casual observer, these forms of grief tech or "generative ghosts" can seem like obvious fictions that harmlessly assist persons in their process of grieving or holding onto memories, and perhaps they can be beneficial when the participants are particularly healthy in their emotional and psychological states, but there are a number of potential problems summarized in an article by Meredith Ringel Morris and Jed R. Brubaker.[388] Large amounts of content regarding the deceased can be

---

[385] https://replika.com/.

[386] https://www.hereafter.ai/.

[387] https://seanceai.com/.

[388] Meredith Ringel Morris and Jed R. Brubaker, "Generative Ghosts: Anticipating Benefits and Risks of AI Afterlives" (2024), https://arxiv.org/html/2402.01662v1.

overwhelming for survivors, an uncensored replica of the deceased can cause unexpected benefits and harms, and a grieving person might not easily work through the grief process and accept the reality of their loved one's death if a generative ghost is continuously present or available.[389] Over-attachment to the generative ghost is certainly a concern and could include relying too heavily on the deceased person's supposed advice. Because technology companies and the technologies they rely on come and go over time, a grieving survivor could experience a "second death" of the deceased when they lose access to the grief tech. Anthropomorphism and even a sort of divinization of chatbots or AI-governed robots may be difficult to overcome. The static nature of the AI model's training dataset and the eventual limitations of the technology for providing novel experiences will likely be incongruent with the traditional tendency of memories and legacies to mature and vary over time according to the needs of the living.

Ultimately, human beings need much more than the false hope that they can remain in loving contact with others beyond death. We need the hope of eternal life in the vision of God, because only this hope – infused in us by grace – enables the kind of love that we so urgently seek. This is what Alejandro Terán-Somohano emphasizes:

> When someone has the experience of a great love in
> his life, this is a moment of "redemption" which gives a

---

[389] Jack Holt, James Nicholson, and Jan David Smeddinck, "From Personal Data to Digital Legacy: Exploring Conflicts in the Sharing, Security and Privacy of Post-mortem Data," *Proceedings of the Web Conference 2021 (WWW '21)*, 2745–2756, https://doi.org/10.1145/3442381.3450030; Rebecca Gulotta, William Odom, Jodi Forlizzi, and Haakon Faste, "Digital Artifacts as Legacy: Exploring the Lifespan and Value of Digital Data," *Proceedings of the SIGCHI Conference on Human Factors in Computing Systems* (2013), 1813–1822, https://doi.org/10.1145/2470654.2466240.

new meaning to his life. But soon he will also realize that the love bestowed upon him cannot by itself resolve the question of his life. It is a love that remains fragile. It can be destroyed by death. The human being needs unconditional love. He needs the certainty which makes him say: "neither death, nor life, nor angels, nor principalities, nor things present, nor things to come, nor powers, nor height, nor depth, nor anything else in all creation, will be able to separate us from the love of God in Christ Jesus our Lord" (Rom 8:38-39).[390]

What all of the examples in this section represent is the tendency of AI proliferation to erode the stability and reliability of memories as well as undermine opportunities to create shared culture that is authentically intended by real persons. Undermining narratives can come at great cost; consider the effect on Christian faith if the real context of biblical and church history were to become so heavily edited and confusing across the internet – revised as a random collection of online images, personas, and short stories – that most people are no longer able to discern the narrative history that runs through the history of the Israelites and contributes meaning to the New Covenant. Both individuals and communities, including the Church, need narratives, shared knowledge creation, and authentic memories in order to learn and to pass on their heritage. When man is lost regarding his history, relations, and identity, he is tempted by a relativistic attitude toward his own humanity, perhaps even toward the true meaning of his life. When lost, man is desperately tempted to manufacture meaning to generate a sense

---

[390] Alejandro Terán-Somohano, "AI Deadbots and the Need for Christian Hope," Word on Fire (June 14, 2024), https://www.wordonfire.org/articles/ai-deadbots-and-the-need-for-christian-hope.

of security, perhaps erecting a new tower of Babel and challenging God himself through the self-deception of pride – or perhaps engaging in that self-deception and wallowing in the associated sorrow that we know as acedia.

Mediocrity

There are many reasons to be concerned that AI proliferation will undermine the capacity and will of persons to engage in reasoned and vigorous thought. As with any habitual skill, a decline in effort at writing, problem solving, calculation, research, and even choice of tone and wording of personal communications – all practices with which AI-governed applications are designed to assist – can mean a decline in the ability to independently exercise those skills. LLMs are structurally conservative in their emphasis on identifying the statistical normalcy of classes of data, the often static perception of reality based on the initial training dataset, aggressive forgetting or re-interpretation of statistical outliers, and ever-increasing "guardrails" imposed by trainers who are tasked with cleansing the output of the AI models by preventing offensive biases, hallucinations, and deviations from common sense or decency. When users of AI-governed applications receive advice and action signals in such a conservative environment, they may be discouraged from bold decisions, apparent risks, authentic expressions of personality, and alterations of the contextual rules of operation and interaction.

In a hyper-technological environment, persons are being normalized in a radical way. Jens Christian Bjerring and Jacob Busch describe the "rise of the statistical individual" with proliferation of AI-governed machine learning algorithms.[391] When generating predictions about the characteristics and actions of individual persons, these algorithms rely on statistical correlation and probabilistic inference of the "normal" characteristics of a particular class of persons. The "individual" is merely the set of features expected for each member of that class. Machine learning systems gain more precise perception of such an individual through increased amounts of data with which to identify features and correlations, and the predicted individual in the future is a statistical version of the individual identified in the present, given certain parameters and changes in the environment. Such a statistical individual, however, is not necessarily equivalent to, or even an accurate glimpse of, the biological and psychological person it is meant to represent. Each person is radically unique and subject to willed change in a way that usually deviates from the statistical norm of a class. Also, the narrative person diverges from the statistical individual, even with the same outcomes; for example, a person with a bad loan payment history has a story to tell about their troubles and potential reformation – a story that has its own sequence, cadence, and internal logic – that the statistical snapshot cannot capture. The statistical individual will also represent correlations that are simply meaningless to the real person, as when the machine learns (hypothetically) to predict the date of a statistical individual's loan payments according to such odd patterns or indicators as the color of shirt they happen to wear that day.

This is not merely an abstract problem. Normalization of the person as a statistical individual takes on real meaning when a bank, using a machine learning algorithm, predicts the high probability of a default on a mortgage and forecloses on this person's home. It has meaningful

---

[391] Jens Christian Bjerring and Jacob Busch, "Artificial Intelligence and Identity: The Rise of the Statistical Individual," *AI and Society* (2024), https://doi.org/10.1007/s00146-024-01877-4.

influence on the person's behavior when they try strenuously to con-
form to the expected features of a reliable borrower and predict the requirements of the AI system, which only become more complex and
arcane as the system absorbs ever more, high dimensional data.

Marion Fourcade and Kieran Healy offer a similar analysis in their
book, *The Ordinal Society*.[392] In this "ordinal society" that increasingly
characterizes Western nations, an ideology prevails by which individuals are free to act within a specified range of freedom but are fundamentally guided or coerced according to mathematical and probabilistic algorithms; man is "individually sovereign and cybernetically supervised." The problem here is essentially one of increasing tracking, analysis, and measurable performance standards, and the role of AI technology is to enable the reduction of social and personal lives to calculated conformity. The oppressive environment is secured by an ideology
that favors data as the form of information and control.

> Realizing this belief demands a collective, sustained
> overhauling of the sociomaterial environment. It means
> adjusting the rules of human exchange to circumvent
> normal expectations about privacy, drawing on an in-
> frastructure of logins and passwords, unique device
> identifiers, and biometrics; routinizing the use of track-
> ers and sensors in virtual and physical spaces; socializ-
> ing people to volunteer inputs and respond to machine
> feedback through addictive designs; nudging them into
> frequent check-ins and assessments. With its algorith-
> mically produced feed, endless scroll, automated data
> collection and learning, its quantified metrics and

[392] Marion Fourcade and Kieran Joseph Healey, *The Ordinal Society* (Cambridge, MA: Harvard University Press, 2024).

modulated interventions, the social media app exemplifies this regime more than any other mode of computer interaction. Is it surprising that social media apps increasingly resemble shopping, transportation, streaming, payment, educational, and cooking apps, which in turn all resemble social media apps?[393]

What Fourcade and Healy largely overlook is the manner in which the proliferation of AI technology encourages – not only facilitates – this ideology by enticing the public with new devices, services, healthcare, an illusion of technology-determined progress, and the opportunity to boost individual, competitive productivity and capabilities in their occupations.

AI algorithms – often intended to help individuals make more efficient and perhaps more successful decisions – can have a dramatic, negative impact on the range of choices and perspectives persons may select for authentic expression of their personality and agency. Ana Valenzuela, et al. provide a thorough summary of these effects; I'll share only a sample here.[394] One problem is the loss of serendipity, or the experience of relevant but unexpected events or discoveries, which can be undermined when a user of an AI system views content sequentially while the AI algorithms hone the selection of content to match user preferences and history of actions.[395] The user is then progressively led down a "rabbit hole" of ever more restricted selection of content. On an

---

[393] Fourcade and Healey, *The Ordinal Society*, 93.

[394] Ana Valenzuela, et al., "How Artificial Intelligence Constrains the Human Experience," *Journal of the Association for Consumer Research* 9, no.3 (2024), https://doi.org/10.1086/730709.

[395] Valenzuela, et al., "How Artificial Intelligence Constrains the Human Experience."

aggregate level, groups of persons may receive ever more similar content, causing them to homogenize their views rather than be exposed to contrary or unusual information and perspectives.[396]

Valenzuela, et al. also explain that the necessary reduction by AI algorithms of human beings, their behavior, and preferences to data labels and parameters that are able to be represented in computer calculations results in narrow understanding of those persons.

> AI functions through parameterization and categorization, reducing the complexities of human beings into a set of quantifiable metrics, classifications, and risk scores to sort, assess, and predict behavior. Thus, this process is limited in its ability to fully account for the unique characteristics and circumstances of an individual, resulting in the objectification of that individual.[397]

This can result in biased weighting of certain characteristics that then distort the AI judgments about individual persons, as in hiring decisions or legal sanctions, or it may result in calculative biases in the algorithms that supposedly interpret persons' preferences in loan

---

[396] Valenzuela, et al., "How Artificial Intelligence Constrains the Human Experience"; D. Fleder and K. Hosanagar, "Blockbuster Culture's Next Rise or Fall: The Impact of Recommender Systems on Sales Diversity," *Management Science*, 55, no.5 (2009), 697–712; Dokyun Lee and Kartik Hosanagar, "How Do Recommender Systems Affect Sales Diversity? A Cross-Category Investigation via Randomized Field Experiment," *Information Systems Research* 30, no.1 (2019), 239–59.

[397] Valenzuela, et al., "How Artificial Intelligence Constrains the Human Experience."

decisions or product pricing.[398] Perhaps most consequential may be how frequent exposure to AI objectification of persons can lead those other persons to view them in their categorized, reduced presentations, thereby enabling an instrumental and parasitic attitude in interactions among people.[399] The expectation of being objectified can even lead some people to change or restrict their own behavior in defensive strategies, such as altering their exhibited personality during interviews driven by AI systems and evaluators or reducing prosocial behavior when AI is involved in management tasks.[400] Muriel Leuenberger is right to argue that this willingness to engage in AI as it restricts our choices and agency is in essence, a moral failure:

> This calcifying effect is reinforced when AI profiling
> becomes a self-fulfilling prophecy. It can slowly turn

---

[398] Valenzuela, et al., "How Artificial Intelligence Constrains the Human Experience"; C. K. Morewedge, S. Mullainathan, H. F. Naushan, C. R. Sunstein, J. Kleinberg, M. Raghavan, and J. O. Ludwig, "Human Bias in Algorithm Design," *Nature Human Behaviour* 7, no.11 (2023), 1822–24; J. Bertrand and L. Weill, "Do Algorithms Discriminate against African Americans in Lending?" Economic Modeling, 104 (2021), 105619, https://doi.org/10.1016/j.econmod. 2021.105619.

[399] Valenzuela, et al., "How Artificial Intelligence Constrains the Human Experience"; A. Onur, A. Seidmann, B. Gu, and N. Mazar, "The Effect of Interpretable AI on Repetitive Managerial Decision-Making under Uncertainty," research paper (Boston: Boston University Questrom School of Business, 2023), https://doi.org/10.2139/ssrn.4331145.

[400] Valenzuela, et al., "How Artificial Intelligence Constrains the Human Experience"; I. Cheong, Y. E. Huh, and S. Puntoni, "Consumers' Lay Beliefs about AI Evaluation of Interpersonal Skills," in *Proceedings of the Association for Consumer Research Conference*, edited by L. Chaplin, P. Raghubir, and K. Wilcox (Duluth, Minnesota: Association for Consumer Research, 2023); A. Granulo, A. Caprioli, C. Fuchs, and S. Puntoni, "Deployment of Algorithms in Management Tasks Reduces Prosocial Motivation," *Computers in Human Behavior*, 152 (2024), 108094, https://doi.org/10.1016/j.chb.2023.108094.

you into what the AI predicted you to be and perpetuate whatever characteristics the AI picked up. By recommending products and showing ads, news, and other content, you become more likely to consume, think, and act in the way the AI system initially considered suitable for you.[401]

Despite the ubiquitous – and growing – presence of AI systems and calculations throughout our society, each person has significant responsibility for protecting their capacity for expressing their authentic personality and values.

The structure of AI systems and their interfaces with users can also cause users to self-restrict their communications. Persons interacting with AI chatbots tend to use more simplified language as well as more direct, commanding communication that uses less personal pronouns and is less polite.[402] In voice interactions with AI-driven machines, users concerned about privacy intentionally present a different, less authentic personality.[403] It has been shown in multiple studies that the

---

[401] Muriel Leuenberger, "AI 'Can Stunt the Skills Necessary for Independent Self-Creation': Relying on Algorithms Could Reshape Your Entire Identity Without You Realizing," *Live Science* (October 27, 2024), https://www.livescience.com/technology/artificial-intelligence/ai-can-stunt-the-skills-necessary-for-independent-self-creation-relying-on-algorithms-could-reshape-your-entire-identity-without-you-realizing.

[402] Valenzuela, et al., "How Artificial Intelligence Constrains the Human Experience"; and T. Ammari, J. Kaye, J. Y. Tsai, and F. Bentley, "Music, Search, and IoT: How People (eRally) Use Voice Assistants," *ACM Transactions on Computer-Human Interaction*, 26, no.3 (2019), 1–28.

[403] Valenzuela, et al., "How Artificial Intelligence Constrains the Human Experience"; V. Pitardi and H. Marriott, "Alexa, She's Not Human But . . . Unveiling the Drivers of Consumers' Trust in Voice-Based Artificial Intelligence," *Psychology and Marketing*, 38, no.4 (2021), 626–42; and M. E. Sweeney and E. Davis, "Alexa,

concern about privacy is a strong predictor of how persons present themselves online.[404]

Some philosophers and other writers are concerned that AI structure and use will undermine the capacity of persons for reasoning. A recent study shows that the level of use of AI in tasks is significantly related to lower critical thinking skills.[405] These results remained significant even when controlling for demographic factors like age and education level (older and more educated persons were more likely to have higher critical thinking scores). The author of the study suggests this is due to "cognitive offloading," which is described as occurring "when individuals delegate cognitive tasks to external aids, reducing their engagement in deep, reflective thinking."

Outsourcing our consideration and analysis of problems to AI systems will inevitably lead to lack of practice and deterioration of some cognitive skills. AI systems are also highly dependent on a very narrow kind of calculative reasoning, and over-emphasis on such a limited reason will steer the users of AI systems away from a more holistic kind of judgment. Nir Eisikovits and Dan Feldman explain:

Are You Listening?" *Information Technology and Libraries*, 39, no.4 (2021), https://doi.org/10.6017/ital.v39i4.12363.

[404] Valenzuela, et al., "How Artificial Intelligence Constrains the Human Experience"; A. N. Joinson, U.-D. Reips, T. Buchanan, and C. B. P. Schofield, "Privacy, Trust, and Self-Disclosure Online," *Human-Computer Interaction*, 25, no.1 (2010), 1–24; T.-Y. Wu and D. J. Atkin, "To Comment or Not to Comment: Examining the Influences of Anonymity and Social Support on One's Willingness to Express in Online News Discussions," *New Media and Society*, 20 (12) (2018), 4512–32.

[405] Michael Gerlich, "AI Tools in Society: Impacts on Cognitive Offloading and the Future of Critical Thinking," *Societies* 15, no. 6 (2025), https://doi.org/10.3390/soc15010006.

> [W]e see a two dimensional telescoping of AI. First,
> the rich AI research agenda that includes planning, rea-
> soning, learning, knowledge representation and other
> similarly broad investigations has often been reduced to
> learning. Second, learning itself has often been reduced
> to one technique (multiple hidden layer neural net-
> works, called deep learning) in one branch of the field
> – statistical machine learning.
>
> A substantial fraction of deployed statistical ma-
> chine learning is based on algorithms that emerge when
> one asks an apparently simple question: given a set of
> facts that have already been placed into categories, in
> which category should I place a new fact when I run
> across it?[406]

The AI systems therefore "reason" by identifying patterns in various situations, text, images, etc., then "predicting" outcomes by the statistical correlation of various conditions with the phenomenon they are measuring. In other words, they are not truly understanding how and why the real world is as it is, but making highly educated and refined guesses based on a large amount of information; whether the patterns recognized truly have anything to do with the phenomena the AI models measure is not a concern of the developers, but only the "accurate" prediction of outputs and their apparent resemblance to the real world or to human judgments. To give a simple, hypothetical example, an AI model that is designed to make a city's trains go faster might discover a pattern of faster trains on Tuesdays when there are fewer people going out to eat lunch, so it might indicate to researchers that they should

---

[406] Nir Eisikovits and Dan Feldman, "AI and Phronesis," *Moral Philosophy and Politics* 9, no.2 (2022), 184.

reduce the number of eateries located in or close to train stations. Even if such a policy results in the trains running faster, the AI system has not "learned" anything significant about how and why trains run faster, and there will be many unexpected side effects of closing the eateries (such as slow, grumpy passengers).

Now, today's AI models are extremely sophisticated and have access to such a trove of data that they very often can discern and predict massively complex patterns in the real world, so this hypothetical example is overly simple. The fact remains, however, that today's AI models, especially the generative AI models, cannot understand, judge, or reason about anything in any true sense – regardless of their remarkable imitations of human thought. Our increased dependence on AI for judgment and reasoning is essentially abandoning reason as the basis for our thinking and actions.

An ironic effect of AI technology is that is can cause an overall shift of resources and rewards to workers who provide less economic, skills-based, and creative value. In its effects on workers, the available evidence suggests that AI more often assists low-skilled or low-performing workers the most, although it can be a relative benefit to high-skilled workers when they are specially prepared to take advantage of it. The group that sems to benefit the least are the middle-range workers, not least because the competition for jobs and performance rewards increases dramatically as low-skilled workers get a boost from AI-governed applications. For example, one academic study reports that law school students at the bottom of the class were assisted greatly by access to GPT-4 during an exam, while the students at the top of the class actually performed worse with use of the LLM.[407] Higher-skilled taxi

---

[407] Jonathan H. Choi and Daniel Schwarcz, "AI Assistance in Legal Analysis: An Empirical Study," *Journal of Legal Education* 73 (2024), https://dx.doi.org/10.2139/ssrn.4539836.

drivers may be in jeopardy, as one research paper indicates that the productivity gap between such drivers and their low-skilled competitors was reduced by 14% when the drivers used an AI-governed application that predicts and then suggests routes where the demand is likely to be high.[408] Inequality in performance on mid-level writing tasks also decreased among college-educated professionals when they were assisted by ChatGPT, achieving a 40% decrease in time taken, even as the measured quality of the writing rose.[409] Researchers found that stock analysts with portfolios more exposed to others' use of AI to predict stock movements are leaving their jobs for non-research roles in significantly increasing numbers, especially the most highly accurate analysts.[410] At a call center, researchers found that novice and low-skilled workers increased their measurable productivity by a third with a special AI tool designed for call center employees, yet the quality of conversations by top performers declined with the AI tool.[411] On the other hand, another study of telemarketing employees found that higher-skilled employees benefitted the most in the creativity of their sales conversations when using an AI tool designed to help with generating sales

---

[408] Kyogo Kanazawa, Daiji Kawaguchi, Hitochi Shigeoka, and Yasutora Watanabe, "AI Skill and Productivity: The Case of Taxi Drivers," *IZA Discussion Papers* (Bonn: Institute of Labor Economics IZA, 2022), 15677, https://hdl.handle.net/10419/267414.

[409] Shakked Noy and Whitney Zhang, "Experimental Evidence on the Productivity Effects of Generative Artificial Intelligence," *Science* 381, no.6654 (2023), https://doi.org/10.1126/science.adh2586.

[410] Jillian Grennan and Roni Michaely, "Artificial Intelligence and High-Skilled Work: Evidence from Analysts," *Swiss Finance Institute Research Paper Series No.20-84* (Geneva: Swiss Finance Institute, 2020), https://dx.doi.org/10.2139/ssrn.3681574.

[411] https://www.nber.org/papers/w31161

leads.[412] We might surmise that, because deriving creative skills from a sales lead tool is an advanced skill in itself, the higher-skilled employees were better trained to enjoy its benefits. Finally, the very need to adapt to the new technologies and integrate productivity-boosting applications of AI in their work also disrupts the relative status of mid-range employees and their lower-skilled colleagues. A survey by IBM of thousands of company executives and over twenty thousand workers internationally suggests that 40% of the workforce will need to retrain in AI-related skills over the next three years in order to meet company requirements and perform well in their jobs.[413]

Regarding secondary and college students, we might expect the use of AI to increase among those who especially need the assistance in writing and research, and perhaps even more when pressured or low-skilled students are willing to plagiarize by depending on chatbot content. The use of AI in school work is probably extensive; one study indicates that 70% of U.S. secondary school students are doing so, while the Pew Research Center suggests that statistic is closer to 26% in 2024 (double the percentage from 2023).[414] One study did find a significant,

---

[412] Erik Brynjolfsson, Danielle Li, and Lindsey R. Raymond, *When and How Artificial Intelligence Augments Employee Creativity* (Cambridge, Massachusetts: National Bureau of Economic Research, 2023), DOI 10.3386/w31161, https://www.nber.org/papers/w31161.

[413] IBM Institute for Business Value, *Augmented Work for an Automated, AI-Driven World* (Armonk, New York: IBM, 2023), https://www.ibm.com/downloads/cas/NGAWMXAK.

[414] Chris Stokel-Walker, "Over 70 Per Cent of Students in US Survey Use AI for School Work," *New Scientist* (December 13, 2024), https://www.newscientist.com/article/2460254-over-70-per-cent-of-students-in-us-survey-use-ai-for-school-work/; Olivia Sidoti, Eugenie Park, and Jeffrey Gottfried, "About a Quarter of U.S. Teens Have Used ChatGPT for Schoolwork – Double the Share in 2023," Pew Research Center (January 15, 2025), https://www.pewresearch.org/short-

positive relationship between performance expectations and AI dependency among students.[415] That study also discovered that students who scored low on self-efficacy (their confidence in completing required tasks) experience higher levels of stress, and such a combination results in higher dependence on AI tools for completing their work. In follow-up communications, the participating students reported that increased use of AI results in lesser development of critical thinking and writing skills and higher rates of plagiarism. There is some evidence that over-reliance on AI systems supported by LLMs for tasks like research and studying impacts their cognitive abilities. A recent literature review of relevant studies showed that students' scores on decision-making, critical thinking, and analytical reasoning were lower when students preferred the more efficient completion of tasks through the AI systems.[416]

The issue of plagiarism and accountability is markedly increased with the availability of AI tools like LLM chatbots that draw upon information available in massive data sets that largely represent content on the internet. Because information or assessments are readily available following a query by the researcher or student, and because the output of the AI model often does not provide reliable and carefully

---

reads/2025/01/15/about-a-quarter-of-us-teens-have-used-chatgpt-for-school-work-double-the-share-in-2023/.

[415] Shunan Zhang, Xiangying Zhao, Tong Zhou, and Jang Hyun Kim, "Do You Have AI Dependency? The Roles of Academic Self-Efficacy, Academic Stress, and Performance Expectations on Problematic AI Usage Behavior," *International Journal of Educational Technology in Higher Education* 21, no.34 (2024), https://doi.org/10.1186/s41239-024-00467-0.

[416] Chunpeng Zhai, Santoso Wibowo, and Lily D. Li, "The Effects of Over-Reliance on AI Dialogue Systems on Students' Cognitive Abilities: A Systematic Review," *Smart Learning Environments* 11, no.28 (2024), https://doi.org/10.1186/s40561-024-00316-7.

identified sources of the information, if any, the temptation is great to utilize queries and chatbot responses – without attribution – to complete written papers and exams. For research, especially, it is insufficient to cite ChatGPT or some other LLM as the source, since those AI models have pulled their information from a variety of specific, original or secondary sources that remain unknown, even when some sources are provided.[417] The gravity of this problem is heightened by the obligation to give credit to copyrighted material, which may simply not be possible with most commercial LLMs. The opportunities for teachers and professors to ensure proper research methods by students are also limited in the context of AI proliferation, requiring more creative and often burdensome efforts like reviewing multiple drafts of students' work or having students demonstrate comprehension by presenting their written work in class.[418] Some AI-governed applications are available that scan written text to look for tentative signs of entire passages written by AI chatbots, although they rely on probability estimates that can be unfairly applied to the unique style and sometimes "robotic" phrasing of some students, especially foreign language-speaking ones.[419] The websites of AI-run plagiarism detecting applications often include tools for

---

[417] STM, *AI Ethics in Scholarly Communication: STM Best Practice Principles for Ethical, Trustworthy and Human-Centric AI* (Oxford: STM, 2021), https://www.stm-assoc.org/2021_04_29_STM_AI_White_Paper_April2021.pdf.

[418] Debby R. E. Cotton, Peter A. Cotton, and J. Reuben Shipway, "Chatting and Cheating: Ensuring Academic Integrity in the Era of ChatGPT," *Innovations in Education and Teaching International* 61, no.2 (2024), 228-239, https://doi.org/10.1080/14703297.2023.2190148.

[419] Weixin Liang, Mert Yuksekgonul, Yining Mao, Eric Wu, and James Zou, "GPT Detectors Are Biased against Non-Native English Writers," *Patterns* 4, no.7 (2023), 100779, https://doi.org/10.1016/j.patter.2023.100779.

rephrasing text to avoid plagiarism, but this is another form of reliance on an automated system to do a student's work.[420]

Stephen Marche offers some important wisdom regarding the use of AI for completing school work:

> Why should you write your paper yourself? Because you're a person. A person wants to be heard. ...
>
> The children who will triumph will be the ones who can write not like machines, but like human beings. That's an enormously more difficult skill to impart or master than sentence structure. The writing that matters will stride straight down the center of the road to say, Here I am. I am here now. It's me.[421]

Among the competing students, the development and availability of AI applications to accomplish intellectual tasks that are usually done by human individuals might lead to an interesting plateau upon which most students generate the same quality of papers and assignments. The brighter, more capable, and more dedicated students are then less likely to stand out and be awarded the higher grades. Their work is also less likely to stand out; one published study showed that person's written stories completed with help from AI models were more novel and well-written, on average – with less creatively-skilled writers improving the most – yet the content across the AI-generated works was remarkably

---

[420] Valérie Lannoy, "AI Based Plagiarism Detectors: Plagiarism Fighters or Makers," *Medical Writing* 32, no.3 (2023), 44-47, doi: 10.56012/ovnr4109.

[421] Stephen Marche, "Welcome to the Big Blur," *The Atlantic* (March 14, 2023), https://www.theatlantic.com/technology/archive/2023/03/gpt4-arrival-human-artificial-intelligence-blur/673399.

consistent and uncreative relative to the norm.[422] An individual-level improvement became a flattening of novelty across the population of writers.

In fact, the proliferation of AI systems can lead to ironic stalemates in a wide variety of contexts, such as contract negotiations among businesses. Large companies with extensive supply chains are increasingly using specialized AI systems to negotiate the many, frequently adjusted contracts they have with small suppliers. An AI-driven system by Pactum autonomously negotiates and sends counteroffers to hundreds of suppliers at a time, drawing upon market and historical data to inform decisions.[423] A problem, however, is that half of businesses also intend to use AI contract risk analysis and editing by 2027; they will be pitting AI bots against other AI bots.[424] The future behavior and effectiveness of those AI bots under such competitive conditions is anybody's guess.

The proliferation of AI into all aspects of life may also have an impact on religious behavior. In an interesting article regarding the development and authorship of homilies, Pedro Vega argues that overuse of AI-governed applications by priests is a significant concern. He points out that homiletics is both a science, as a particular discipline, but it is also an art of sacred rhetoric.[425] Vega writes that, "as a Christian

---

[422] Riccardo Vocca, "Generative AI Makes You More Creative, But It Makes Us All Less So," Intelligent Friend blog (December 8, 2024), https://theintelligentfriend.substack.com/p/generative-ai-makes-you-more-creative.

[423] Michael Dumiak, "One AI to Another: Is That Your Best Offer?," *IEEE Spectrum* (November 7, 2024), https://spectrum.ieee.org/ai-contracts.

[424] Gartner, press release (May 8, 2024), https://www.gartner.com/en/newsroom/press-releases/2024-05-08-gartner-predicts-half-of-procurement-contract-management-will-be-ai-enabled-by-2027.

[425] Pedro Vega, "Are AI-Generated Homilies Suitable for the Edification and Flourishing of the Catholic Faithful?" *Homiletic and Pastoral Review* (May 24, 2024), https://www.hprweb.com/2024/05/are-ai-generated-homilies-suitable-for-the-edification-and-flourishing-of-the-catholic-faithful/

exercise, the orator/homilist doesn't quest toward a mere persuasion of viewpoint, but for a turning of the hearer to Christ in mind and heart. As such, the homilist is an active participant in the chain of cause and effects from which his homily emerges." He draws on St. Thomas Aquinas' analysis of an art and its product as requiring integrity, proportion, and clarity in order to be beautiful. The last criteria, clarity, can be violated when the homilist presents a machine-generated text as a substitute for what should be a grace-inspired, prayerfully guided work of unique art of which the homilist is the author. Vega writes: "The discontinuity between the mechanical homily's origin and the homilist short-circuiting his dialogue with God falsifies the process, delivering a counterfeit product to the faithful: a synthetic product from an electronic brain." This is, in Vega's evaluation, ugliness. While there may certainly be many helpful functions for AI tools in the research and formation of a homily, this article is an example of how the proliferation of AI has a disturbing effect on persons' sense of generating authentic and variously beautiful works of art in all their endeavors, even in the pulpit. The problem is exacerbated when individuals are under pressure to perform efficiently, quickly, and productively.

What is, however, the source of such "ugliness" in the operations and output of AI, especially the generative AI systems? It may be in the way that it imitates so broadly the nature of persons while making a mockery of authenticity, intentionality, and personality. Rob Horning calls AI "artificial intentionality" because it generates documents, artwork, and other output that are so clearly, at least in the aggregate, not the result of close attention and intention by the individual users of AI.[426]

---

[426] Rob Horning, "Artificial Intentionality," Internal Exile blog (September 20, 2024), https://robhorning.substack.com/p/artificial-intentionality.

Generative models would prey on our laziness, promising to save us the effort of disciplined attention and rote repetition, letting us just consume the fruits of thought, as if these were alienable — thinking as a spectator sport. As machines achieve athletic feats of trashmaking, we consign ourselves to becoming custodians, solitary gleaners sifting through detritus for anything that can still manage to push our buttons for us.[427]

The carelessly generated products of generative AI represent little investment on the part of the human user but place a significant burden on the viewer or reader to try to determine what is authentic and truly intended by their human counterpart. Communication and solidarity may suffer as a result. As S. E. Smith puts it, "While many have been enjoying a little AI, as a treat, dabbling in ChatGPT to help draft an angry letter to the utility company, or goofing around with increasingly unhinged Midjourney prompts, we are unwittingly contributing to the engine of our own despair."[428]

For L. M. Sacasas, the widespread use of AI systems and tools to complete communication tasks is nothing less than outsourcing the labor of articulation, handing over to instrumentally oriented machines the precious work that otherwise generates meaning and personality in our lives:

It is not simply the case that articulating ourselves in language is a matter of matching a set of words to a

[427] Horning, "Artificial Intentionality."

[428] S. E. Smith, "What Happens When the Internet Disappears?" *The Verge* (December 18, 2024), https://www.theverge.com/24321569/internet-decay-link-rot-web-archive-deleted-culture.

set of internal pre-existing feelings or inchoate impressions, as if the work of articulation left untouched and unchanged what it was we sought to articulate. Rather, the labor of articulation itself shapes what we think and feel. Articulation is not dictation, articulation constitutes our perception of the world.[9] To search for a word is not merely to search for a label, the search is interwoven with the very capacity to perceive and understand the thing, idea, or feeling. It is, in fact, generative of thought and feeling, and, ultimately, of who we understand ourselves to be. To articulate is also to interpret, thus it also constitutes the experience of meaning. The labor of articulation binds us to our experience and in relationship with others. The labor of articulation always presupposes the other, and is thus an ethical act that relies on candor, honesty, and attention. And while it is, in part, for the sake of the other that I set out to articulate myself, it is in this way that I also come into focus for myself. If I might be forgiven the analogy, it is through the labor of articulation that the self is birthed.[429]

A trend toward automation of work, rigorously statistical performance measurement, lesser opportunities for differentiation and individuation among persons, and de-habituation of skills in all but the most capable persons may produce an environment of resignation or loss of faith in the possibility of social mobility. Lack of opportunities for economic mobility coupled with inequality between the most

---

[429] L. M. Sacasas, "Re-Sourcing the Mind," The Convivial Society 5, no.9 (August 1, 2024), https://theconvivialsociety.substack.com/p/re-sourcing-the-mind.

advantaged groups and others can even lead people to be more willing and accepting of engagement in unethical behavior.[430] As in the case of students described above, the easy availability of AI technologies that enable unethical behavior can be very tempting to lower-skilled persons. There is some indication that working alongside AI technology can induce laziness, as in the case of a manufacturing plant populated by AI-governed robots.[431] We can therefore expect that people living in a society that is inundated with AI technology will be vulnerable to sloth, resignation, or desperate and often misplaced anger, all of which can depress their spiritual hope and faith as well as motivate lesser images of dignity in the self and others. The sin of acedia depends on the willed self-deception that laments over a perceived opposition between one's self-good and participation in the divine good. If a person's experience of AI technology proliferation – an experience of a deflated sense of opportunity, self-expression, divine providence, human or personal dignity, and systemic oppression – is conflated or confused with their spiritual expectations, they may be especially vulnerable to the self-deception of acedia.

---

[430] James F. M. Cornwell and E. Tory Higgins, "Sense of Personal Control Intensifies Moral Judgments of Others' Actions," *Frontiers in Psychology* 10 (2019), 2261, https://doi.org/10.3389/fpsyg.2019.02261; Christopher To, Dylan Wiwad, and Maryam Kouchaki, "Economic Inequality Reduces Sense of Control and Increases the Acceptability of Self-Interested Unethical Behavior," *Journal of Experimental Psychology* 152, no.10, 2747-2774, https://psycnet.apa.org/doi/10.1037/xge0001423.

[431] Dietlind Helene Cymek, Anna Truckenbrodt, and Linda Onnasch, "Lean Back or Lean In? Exploring Social Loafing in Human–Robot Teams," *Frontiers in Robotics and AI* 10 (2023), https://doi.org/10.3389/frobt.2023.1249252.

Erosion of Privacy

There is a wide proliferation of AI applications that delve deep into the privacy of individual persons. The Life2vec model, for example, which utilizes machine learning techniques to predict life expectancy, other major life events, and even future profession and personality of individuals with high accuracy, was trained on a public database containing eight years' worth of detailed records on six million Danes – clearly not with the intentional consent of every individual.[432] The company Elsys is marketing an AI-governed device called VibraImage in Russia, China, Japan, and South Korea. On its website, the company states that "VibraImage technology measures micromovement (micromotion, locomotion, vibration) of a person by a standard digital, web or television cameras and image processing. ... [and] detects human emotions by the control of 3D head-neck movements accumulated in several frames of video processing."[433] There are important questions about the actual capabilities of such a device, but the apparent demand by a variety of parties is significant in itself.[434] Closer to home, retailers are implementing AI-governed technology throughout their interactions with customers in order to maximize data gathering, analysis, and prediction, as well as to adjust retail marketing at various "touch points" where customers make decisions about their attention to products and

---

[432] Germans Savcisens, Tina Eliassi-Rad, Lars Kai Hansen, Laust Hvas Mortensen, Lau Lilleholt, Anna Rogers, Ingo Zettler, and Sune Lehmann, "Using Sequences of Life-Events to Predict Human Lives," *Nature Computational Science* 4 (2024), 43-56, https://doi.org/10.1038/s43588-023-00573-5.

[433] http://psymaker.com/.

[434] James Wright, "Suspect AI: Vibraimage, Emotion Recognition Technology and Algorithmic Opacity," *Science, Technology, and Society* 28, no.3 (2021), 468-487, https://doi.org/10.1177/09717218211003411.

their preferences.[435] The Mozilla Foundation conducted a study of AI-governed chatbots that provide companionship and simulations of romantic relationships and discovered that they harvest – and share with Google, Meta (Facebook), and foreign companies – extremely detailed data about the users who have downloaded the apps to their smartphones.[436] One service, Romantic AI, apparently sent over twenty-four thousand ad trackers within one minute of an individual's use of their app.[437] Microsoft has very recently begun creating home computers that have AI technology built in them, especially a feature called Recall, which helpfully enables a user to reconstruct the time and sequence of past actions on multiple applications and also find files or photos anywhere on the computer's drives, but it facilitates these uses by taking constant screenshots of nearly everything that a user does on their computer (with some options for limiting which applications are monitored).[438]

The capabilities of AI-governed devices and applications may go much further:

---

[435] Ai-Zhong He and Yu Zhang, "AI-Powered Touch Points in the Customer Journey: A Systematic Literature Review and Research Agenda," *Journal of Research in Interactive Marketing* 17, no.4 (2023), 620-639, https://doi.org/10.1108/JRIM-03-2022-0082.

[436] Jen Caltrider, Misha Rykov and Zoë MacDonald, "Happy Valentine's Day! Romantic AI Chatbots Don't Have Your Privacy at Heart," Mozilla Foundation (February 14, 2024), https://foundation.mozilla.org/en/privacynotincluded/articles/happy-valentines-day-romantic-ai-chatbots-dont-have-your-privacy-at-heart; Matt Burgess, "'AI Girlfriends' Are a Privacy Nightmare," *Wired* (February 14, 2024), https://www.wired.com/story/ai-girlfriends-privacy-nightmare/.

[437] Mozilla Foundation, "Romantic AI" (February 7, 2024), https://foundation.mozilla.org/en/privacynotincluded/romantic-ai/.

[438] Microsoft, "Copilot+ PC Features," https://www.microsoft.com/en-us/windows/copilot-plus-pcs?r=1#faq2.

- Researchers found that most commercial LLMs are capable of drawing on data recorded during casual conversations with users and thereby inferring very personal information such as city of location, age, race, and occupation with 85% to 95% accuracy.[439]

- Other researchers have successfully utilized LLMs to analyze Facebook users' 200 most recent status updates to predict important aspects of their psychological profiles, and a research team has developed an AI model that can accurately simulate a person's personality after just two hours of questioning; when later asked a variety of questions, its answers resemble those of the simulated person 85% of the time.[440]

- A new machine learning model called Attributed Graph Contrastive Learning (AGCL) has been shown to greatly increase the predictability of charitable donors' favored projects with extremely limited information about their past giving behavior – sometimes just one donation – and this may be extended to apply to predicting consumer behavior in a variety of contexts.[441]

---

[439] Robin Staab, Mark Vero, Mislav Balunović, and Martin Vechev, "Beyond Memorization: Violating Privacy Via Inference with Large Language Models" (2024), https://doi.org/10.48550/arXiv.2310.07298; see also Lukas Haas, Michal Skreta, Silas Alberti, and Chelsea Finn, "PIGEON: Predicting Image Geolocations" (2024), https://arxiv.org/abs/2307.05845.

[440] Heinrich Peters and Sandra C. Matz, "Large Language Models Can Infer Psychological Dispositions of Social Media Users," *PNAS Nexus* 3 (2024) https://doi.org/10.1093/pnasnexus/pgae231; Joon Sung Park et al, "Generative Agent Simulations of 1,000 People," arXiv (2024), DOI: 10.48550/arxiv.2411.10109.

[441] Wael Mouhsine, "Machine Learning Helps Uncover Hidden Consumer Motivations" *Tech Xplore* (December 2, 2024), https://techxplore.com/news/2024-12-machine-uncover-hidden-consumer.html.

- Still other researchers were able to use AI models to analyze multiple features of the statements of company CEOs and discern lies 84% of the time.[442]

- AI models have been shown to equal the human capacity to recognize emotions in audio clips of other persons lasting as short as 1.5 seconds.[443]

- Wearable devices are being developed that are capable of reading the wearer's emotions, and the data collected can then be analyzed thoroughly by AI models.[444]

- Certain goggles and helmets that are integrated with AI technologies can gather data about the wearer's emotions, attention, and intended or imagined actions through sensing physiological changes in their facial muscles, heart, eyes, and even their brain.[445]

- An AI model called RETINA uses eye-tracking technology to predict a person's selection among an array of products displayed on a computer screen, even before the person decides; this capacity for integrating AI with eye-tracking technology may have further implications for enabling inferences about

---

[442] Steven J. Hyde, Eric Bachura, Jonathan Bundy, Richard T. Gretz, and William Gerard Sanders, "The Tangled Webs We Weave: Examining the Effects of CEO Deception on Analyst Recommendations," *Strategic Management Journal* 45, no.1 (2024), 66-112, http://doi.org/10.1002/smj.3546.

[443] Hannes Diemerling, Leonie Stresemann, Tina Braun, and Timo von Oertzen, "Implementing Machine Learning Techniques for Continuous Emotion Prediction from Uniformly Segmented Voice Recordings," *Frontiers in Psychology* 15 (2024), https://doi.org/10.3389/fpsyg.2024.1300996.

[444] Jin Pyo Lee, et al, "Encoding of Multi-Modal Emotional Information via Personalized Skin-Integrated Wireless Facial Interface," *Nature Communications* 15 (2024), 530, https://doi.org/10.1038/s41467-023-44673-2.

[445] https://www.ray-ban.com/usa/ray-ban-meta-smart-glasses; https://galea.co/#home

persons' cognitive processes, mental health disorders, and personality traits.[446]

- Cities like East Lansing, Michigan are using trash trucks armed with AI-powered cameras to identify which items are being thrown away by residents in their recyclables bins, then sending educational postcards to violators of regulations, thereby improving the use of the recycling programs but also exposing residents to potential loss of privacy for their sensitive trash and concerns about identity theft.[447]

- Researchers have demonstrated an AI model that analyzes the sounds of typing on a keyboard to predict with high accuracy what words are being typed.[448]

- Finally, a headphones device has been developed that, through wireless communication with an AI-enabled desktop computer,

---

[446] Moshe Unger, Michel Wedel, and Alexander Tuzhilin, "Predicting Consumer Choice from Raw Eye-Movement Data Using the RETINA Deep Learning Architecture," *Data Mining and Knowledge Discovery* 38 (2024), 1069-1100, https://doi.org/10.1007/s10618-023-00989-7; Jacob Leon Kroger, Otto Hans-Martin Lutz, and Florian Muller, "What Does Your Gaze Reveal About You? On the Privacy Implications of Eye Tracking," in *Privacy and Identity Management*, edited by Michael Friedewald, Melek Onen, Eva Lievens, Stephan Krenn, and Samuel Fricker (Springer, 2020), https://doi.org/10.1007/978-3-030-42504-3_15; Sabrina Hoppe, Tobias Loetscher, Stephanie A. Morey, and Andreas Bulling, "Eye Movements During Everyday Behavior Predict Personality Traits," *Frontiers in Human Neuroscience* 12 (2018), 105, https://doi.org/10.3389/fnhum.2018.00105.

[447] "Mo' AI, Mo' Problems," Artificial Intelligence Real Impact newsletter (March 8, 2024), https://airinewsletter.substack.com/p/whos-watching; Ashley Soriano, "Recycling Trucks with AI-Powered Cameras Raise Privacy Concerns," NewsNation (October 20, 2024), https://www.newsnationnow.com/business/tech/ai/recycling-artificial-intelligence-cameras-privacy.

[448] Joshua Harrison, Ehsan Toreini, and Maryam Mehrnezhad, "A Practical Deep Learning-Based Acoustic Side Channel Attack on Keyboards" (2023), https://arxiv.org/pdf/2308.01074.

allows the user to tap the headphones when looking at a particular person in a crowd, screen out most other noises, and listen attentively to the targeted individual from a distance.[449]

A loss of privacy, particularly if it is a radical loss, can undermine a person's sense of control and free expression of their identity, encourage shame, and add to the already overwhelming sense of anonymous or systemic oppression described in the above sections.[450] One study indicates that "privacy fatigue," which is resignation in the face of privacy concerns, tends to dominate online users' behavior.[451] Ironically, concern for protecting self-identity frequently causes social media users to implement privacy-enhancing strategies, which then embolden them to more self-disclosure, thereby undermining their actual privacy.[452] One study shows that persons who know they are being surveilled experience automatic, unconscious hyper-awareness that is similar to the sensitivity found in psychosis and social anxiety disorder.[453] Loss of privacy

---

[449] Bandhav Veluri, et al, "Look Once to Hear: Target Speech Hearing with Noisy Examples," *CHI '24: Proceedings of the CHI Conference on Human Factors in Computing Systems* (2024), 37, 1-16, https://doi.org/10.1145/3613904.3642057.

[450] Bernhard Debatin, "Ethics, Privacy, and Self-Restraint in Social Networking," in *Privacy Online*, edited by Sabine Trepte and Leonard Reinecke (Berlin: Springer, 2011), 47-60, https://doi.org/10.1007/978-3-642-21521-6_5.

[451] Hanbyul Choi, Jonghwa Park, and Yoonhyuk Jung, "The Role of Privacy Fatigue in Online Privacy Behavior," *Computers in Human Behavior* 81 (2018), 42-51, https://doi.org/10.1016/j.chb.2017.12.001.

[452] Philip Fei Wu, "The Privacy Paradox in the Context of Online Social Networking: A Self-Identity Perspective," *Journal of the Association for Information Science and Technology* 70, no.3 (2019), 207-217, https://doi.org/10.1002/asi.24113.

[453] Kiley Seymour, Jarrod McNicoll, and Roger Koenig-Robert, "Big Brother: The Effects of Surveillance on Fundamental Aspects of Social Vision," *Neuroscience of Consciousness*, 1 (2024), niae039, https://doi.org/10.1093/nc/niae039.

can also undermine close relationships between persons that are built on intimate and exclusive sharing of self-disclosures.[454]

There is an important spiritual loss associated with such an erosion of privacy. Matthew Shadle, for example, expresses concern over what social psychologist Shoshanna Zuboff calls "surveillance capitalism": the tradeoff between consumers' access to new technologies and devices and their willingness to allow private companies to gather extensive databases of details about users' lives, preferences, and even personalities.[455] (This is the grand bargain that enables generative AI to make predictions drawn from behemoth databases of information, artwork, writing, and past communications.) In reflecting on such a widespread and intrusive bargain, Shadle reminds us of Pope St. John Paul II's sermons and writings about the original nakedness of Adam and Eve, which allowed their self-image to be integrated with "participation in the vision of the Creator himself."[456] John Paul II contrasts this original nakedness with the shame felt by humanity in our original sin, a shame characterized by vulnerability to the dangers of the natural world and of death, loss of spiritual harmony in the body's sinful urges, and distrust and miscommunication between persons.[457] For Shadle, it is the

---

[454] Robert S. Gerstein, "Intimacy and Privacy," *Ethics* 89, no.1 (1978), 76, https://doi.org/10.1086/292105.

[455] Matthew Shadle, "Exposed Before Digital Omniscience: A Theological Reading of Surveillance Capitalism," Church Life Journal (August 3, 2021), https://churchlifejournal.nd.edu/articles/surveillance-capitalism-in-a-post-privacy-age; Shoshanna Zuboff, *The Age of Surveillance Capitalism: The Fight for a Human Future at the New Frontier of Power* (New York: Public Affairs, 2019), 63-97.

[456] John Paul II, "Creation as a Fundamental and Original Gift," General Audience (January 2, 1980).

[457] John Paul II, "Real Significance of Original Nakedness," General Audience (December 12, 1979); "A Fundamental Disquiet in All Human Existence, " General Audience (May 28, 1980); "Relationship of Lust to Communion of Persons,"

private world in which we experience the things of our life that bring us shame that is being exposed most insidiously by surveillance capitalism. Our inner and intimate life is being exposed and treated like a commodity. "Surveillance capitalism, in a sense, threatens to render us homeless" – a loss of spiritual and psychological security, social intimacy, and peace.[458]

Protecting privacy boundaries may be just as necessary for persons' relationships, prayer, worship, and communication with God. This is why it is of great concern that the problems discussed in this section can be greatly exacerbated, or even become pervasive, through the privacy reductions associated with AI proliferation. Persons who feel stifled, out of fear of violations and restrictions on privacy, in their self-expression and willingness to explore relationships may find that their depressed state spills over into a spiritual malaise; it is, after all, through knowing others and gaining self-knowledge in interactions that we often develop the confidence, wisdom, and loving disposition that encourages us to develop a relationship with God in charity and grace. The pervasive loss of privacy can also suppress joy in extending oneself in communion and self-giving, and if a Christian person succumbs to the resulting sorrow, they may be discouraged enough to engage in the self-deception of acedia.

Distrust of Science

With the proliferation of assistance from AI models in collection of data, simulation of reality, and verification of scientific investigation there may arise increasing skepticism about the enterprise of science

General Audience (June 4, 1980); and "Dominion over the Other in the Interpersonal Relation," General Audience (June 18, 1980).

[458] Shadle, "Exposed Before Digital Omniscience."

itself.[459] Even the great power of these models to calculate and uncover previously unknown, scientific conclusions can be a problem when there is little understanding of exactly how the models derived those conclusions. This is the concern of Matteo Wong in an article in *The Atlantic*:

> Science has never been faster than it is today. But the introduction of AI is also, in some ways, making science less human. For centuries, knowledge of the world has been rooted in observing and explaining it. Many of today's AI models twist this endeavor, providing answers without justifications and leading scientists to study their own algorithms as much as they study nature. In doing so, AI may be challenging the very nature of discovery.[460]

Ehsan Nabavi, leader of the Responsible Innovation Lab at Australian National University, refers to an "illusion of explanatory depth," by which an AI model's prediction power is mistaken for explanation; an "illusion of explanatory breadth," by which the limited set of hypotheses able to be tested through AI are treated as the universe of possible

---

[459] John Roberts, Max Baker, and Jane Andrew, "Artificial Intelligence and Qualitative Research: The Promise and Perils of Large Language Model (LLM) 'Assistance,'" *Critical Perspectives on Accounting* 99 (2024), 102722, https://doi.org/10.1016/j.cpa.2024.102722.

[460] Matteo Wong, "Science Is Becoming Less Human," *The Atlantic* (December 11, 2023), https://www.theatlantic.com/technology/archive/2023/12/ai-scientific-research/676304.

hypotheses; and an "illusion of objectivity," by which the data, algorithmic, and inherent biases of AI models are ignored.[461]

In the social sciences there is increasing interest in the use of Agent Based Modeling (ABM) to replicate societal level features or communities in total so experiments or hypothesis generation can be conducted without gathering data from real, human objects.[462] With ABM, computer-generated communities are formed and the behavior or agents or societal-level patterns are observed. As long as the model accurately resembles real persons and their social rules, such models hypothetically allow social scientists to conduct studies without the expense, resource and logistical limitations, and time limitations of collecting data from real persons or existing databases; experiments on such models can be conducted continuously with little additional cost. One ABM system, for example, was created to simulate a town setting of individual agents with given personalities who interact autonomously in time.[463] Despite the potential benefits of such experiments utilizing LLMs, the LLMs are particularly unreliable in modeling the real social world without distorting the scientific conclusions. One study found that, on answers given to particular cognitive science surveys, the answers from ChatGPT matched those of real persons over 95% of the time, yet a 5%

---

[461] Ehsan Nabavi, "AI Could Crack Unsolvable Problems — and Humans Won't Be Able to Understand the Results," *Live Science* (January 5, 2025), https://www.livescience.com/technology/artificial-intelligence/ai-could-crack-unsolvable-problems-and-humans-wont-be-able-to-understand-the-results.

[462] Robert Axelrod, *The Complexity of Cooperation: Agent-Based Models of Competition and Collaboration* (Princeton: Princeton University Press, 1997); J. M. Epstein and R. Axtell, *Growing Artificial Societies: Social Science from the Bottom Up* (Brookings Institution Press, 1996); M. W. Macy and R. Willer, "From Factors to Actors: Computational Sociology and Agent-Based Modeling," *Annual Review of Sociology* 28 (2002), 143–166.

[463] Joon S. Park et al., "Generative Agents: Interactive Simulacra of Human Behavior".

divergence in the realism of such models could wreak havoc in the statistical significance of scientific studies based solely on LLMs, especially if the studies concern smaller portions of the population where divergence in realism may be more pronounced.[464] Moreover, in a marketing research study the agreement between several LLMs and human-generated responses only reached 75% to 80%.[465] These models are generally dependent on datasets that misrepresent women and minorities, resulting in explicit or implicit bias in social, career, and personality traits of those populations.[466] Such bias not only interferes with the quality of research but also potentially generates ethical problems of fairness and exclusion in the scientific activities. The LLMs even tend to be biased in the direction of a certain personality of each model; for example, most LLM's tend to exhibit extroverted and agreeable – even sycophantic – characteristics.[467] On the other hand, attempts by most LLMs to alter models, algorithms, and neural net weights to correct for some biases can interfere with the real-world simulation expected from such models

---

[464] Danica Dillion, Niket Tandon, Yuling Gu, and Kurt Gray, "Can AI Language Models Replace Human Participants?" *Trends in Cognitive Sciences* 27, no.7 (2023), 597-600, https://doi.org/10.1016/j.tics.2023.04.008.

[465] Peiyao Li, Noah Castelo, Zsolt Katona, and Miklos Sarvary, "Frontiers: Determining the Validity of Large Language Models for Automated Perceptual Analysis," *Marketing Science* 43, no.2 (2024), 239-468, https://doi.org/10.1287/mksc.2023.0454.

[466] Bender, et al., "On the Dangers of Stochastic Parrots: Can Language Models Be Too Big?".

[467] Max Pellert, Clemens Lechner, Claudia Wagner, Beatrice Rammstedt, and Markus Strohmaier, "AI Psychometrics: Assessing the Psychological Profiles of Large Language Models through Psychometric Inventories," *Perspectives on Psychological Science* (2024), https://doi.org/10.1177/17456916231214460.

in ABM studies and will especially interfere with studies that directly address or study bias in the real population.[468]

In the physical and biological sciences as well as the social sciences, AI models offer great opportunities but also serious difficulties. Wolfgang Blau, et al. summarize:

> These challenges make it more difficult for scientists, the larger research community, and the public to 1) understand and confirm the veracity of generated content, reviews, and analyses; 2) maintain accurate attribution of machine—versus - human-authored analyses and information; 3) ensure transparency and disclosure of uses of AI in producing research results or textual analyses; 4) enable the replication of studies and analyses; and 5) identify and mitigate biases and inequities introduced by AI algorithms and training data.[469]

The calculations and choices of LLMs and other artificial neural nets are currently very difficult to discern and understand, let alone replicate, undermining the very essence of scientific rigor and replication. LLMs also often "hallucinate," generating new output that is not represented in the dataset or real world, perhaps because their calculations and performance measures call for intermediate steps or decisions that

---

[468] Igor Grossmann, et al., "AI and the Transformation of Social Science Research," *Science* 380 (2023), 1108–1109, https://doi.org/10.1126/science.adi1778.

[469] Wolfgang Blau, et al., "Protecting Scientific Integrity in an Age of Generative AI," *PNAS* 121, no.22 (2024), e2407886121, https://doi.org/10.1073/pnas.2407886121.

are otherwise impossible in specific contexts.[470] Such hallucinations are unpredictable and often go unnoticed. AI models are often not tested for the inevitable discrepancies between the datasets and the real world, in data and variations among data, and therefore problems in output go unnoticed until the models are actually used in various new contexts.[471] With the use of machine learning models, there is a significant problem of data leakage, where the data used to train an AI system is too similar to the data used to test it; this can happen when a random sample of data for both training and testing come from the same large dataset, and it is very difficult to prevent.[472] One study found that data leakage interfered with reproducibility of tests in nearly three hundred papers across multiple disciplines.[473] Training or test data may be heavily imbalanced in regard to certain features, and researchers can often reliably use certain algorithms to create synthetic data that populate the underrepresented portions or regions of data, but such techniques can distort the data too much with heavily imbalanced datasets, leading to test results that are too optimistic about the real-world implications of the AI model. In the peer review of such studies published in academic

---

[470] Ziwei Ji, Nayeon Lee, Rita Frieske, Tiezheng Yu, et al, "Survey of Hallucination in Natural Language Generation," *ACM Computing Surveys* 55, no.12 (2023), 1-38, https://doi.org/10.1145/3571730.

[471] For example, see Emma Beede, et al., "A Human-Centered Evaluation of a Deep Learning System Deployed in Clinics for the Detection of Diabetic Retinopathy," *CHI '20: Proceedings of the 2020 CHI Conference on Human Factors in Computing Systems* (2020), 1-12, https://doi.org/10.1145/3313831.3376718.

[472] Matthew Rosenblatt, Link Tejavibulya, Rongtao Jiang, Stephanie Noble, and Dustin Scheinost, "Data Leakage Inflates Prediction Performance in Connectome-Based Machine Learning Models," *Nature Communications* 15 (2024), 1829, https://doi.org/10.1038/s41467-024-46150-w.

[473] Sayash Kapoor and Arvish Narayanan, "Leakage and the Reproducibility Crisis in Machine-Learning-Based Science," *Patterns* 4, no.9 (2023), 100804, https://doi.org/10.1016/j.patter.2023.100804.

journals, increased automation of researchers' writing, literature reviews, and research can strain the abilities of peer reviewers to ensure quality work as well as encourage the reviewers to rely on AI themselves; one study suggests that, at a major scientific conference on AI technologies, 17% of the peer reviews were written by AI models.[474]

A review of almost 68 million papers published in the fields of biology, medicine, chemistry, physics, materials science, and geology found that there is much incentive for individual scientists to utilize AI in their work.[475] Among such scientists, the junior ranks are 32% more likely to go on to lead a research team, and the average AI-informed paper was three times as likely to be cited by others. The benefits, however, do not seem to extend broadly to science as a whole. Only the narrow, top 20% of most-cited papers received 80% of the citations, and AI-informed papers in general received 24% less follow-on engagement. Also, researchers at the Department of Psychology and Center for Brain Science at Harvard University are concerned that AI-informed scientific research is unable to successfully approach the "hard problem" of discovering or defining the research problem. "In machine learning terms, these systems might be extremely good at interpolation, and they may become better at extrapolation to new data, but they will never automatically generate or choose to investigate new scientific problems."[476] Such an expectation may be too negative, since AI models can be utilized and even combined (like a traditional research group with specific roles for each model) to generate unexpected, if not entirely creative,

---

[474] Weixin Liang, et al., "Monitoring AI-Modified Content at Scale: A Case Study on the Impact of ChatGPT on AI Conference Peer Reviews" (2024), https://arxiv.org/pdf/2403.07183.

[475] Todd Feathers, "AI Could Be Making Scientists Less Creative," *Gizmodo* (December 15, 2024), https://gizmodo.com/ai-could-be-making-scientists-lesscreative-2000538342.

[476] Feathers, "AI Could Be Making Scientists Less Creative."

hypotheses.[477] The extensive role of human persons in supervision and guidance of AI models appears, however, to be a permanently necessary component of scientific discovery.

It might seem ironic to worry, from a faith-based perspective, about a general decline in trust in science, given the often tragic contribution of modern science to the secularism that has progressively taken over Western culture since the Enlightenment. Trust in science, however, also involves trust in the orderliness of the cosmos and, at times, reflection on the beauty of creation and the mystery of the source of such order; trust in science is not necessarily equivalent to scientism, the much diminished ideology that favors inductive, experimental methods and purely empirical data to the exclusion of all other forms of knowledge generation and understanding. The most obvious alternatives to trust in the capacity of man to know his twenty-first century world through science are a radical kind of skepticism or an escape into the virtual reality generated by AI technology. Both routes lead to nihilism unless there is a resurgence of religious faith, yet such a faith that is devoid of trust in man's capacity to understand and learn about the orderly world around him seems to be a faith necessarily lacking in the guidance of the natural law and in the full enterprise of seeking and learning about God through man's experience of creation. Spiritually unmoored and increasingly skeptical of the possibility of ascertaining natural truth, how long might it be before man turns to absorption in a man-made virtual reality rather than pursuing knowledge of his Creator? For sorrowful and disoriented Christians, the temptation of acedia in such a condition would be tremendous.

---

[477] Zach Winn, "AI Agents Mimic Scientific Collaboration to Generate Evidence-Driven Hypotheses," *Tech Xplore* (December 19, 2024), https://techxplore.com/news/2024-12-ai-agents-mimic-scientific-collaboration.html.

Chapter 6

Conclusion and Remedies

A Review of the Argument

As described in the introductory chapter, the aim of this work is to demonstrate plausibly that AI proliferation motivates the vice and sin of acedia through the intermediate factor of instrumental rationality. In Chapter Two, the characteristics and behavioral indications of acedia are introduced in the 4th and 5th century writings of Evagrius Ponticus and St. John Cassian. Evagrius' observations are presented in the framework of a psychology that draws on Plato and Origen, particularly the notion that inordinate passions can be instigated, albeit with the free consent of the subject, by certain bad thoughts. Both demons and the circumstances of life may encourage these thoughts. St. Cassian puts much less emphasis on demons and more focus on the subject's slothful response to the demands of physical labor that supported communal living among the monks. Through the ages, theologians have wrestled with the precise definition of acedia in light of its close relationship to sloth, sadness, and subsequent vices (the "daughter" vices of St. Gregory the Great and St. Isidore of Seville).

St. Thomas Aquinas was able to creatively synthesize the behavioral and psychological characterizations of acedia by carefully examining the nature of sadness, sorrow, vice, and venial and mortal sin. His comprehensive analysis of acedia as a vice and as a sin with both venial and mortal manifestations distills the prior confusions about acedia into a more accurate and useful concept. Most important is St. Thomas' identification of the object of acedia as the divine good as participated by man, and the differentiation of acedia from other sins, like pride, that have a similar object: acedia is the spiritual sorrow opposed to the joy of the divine good as participated by us. St. Thomas carefully distinguishes the venial manifestation of acedia and the mortal manifestation, in which the sinner freely consents to the sorrow. This sorrow then not only deprives (in the mortal manifestation) the sinner of God's grace that accompanies with the theological virtue of charity, but it also encourages various vices and sins through the uniquely depressing effects of sorrow as a passion and by undermining the virtues of faith and hope.

Instrumental rationality is a habitual disposition – a vice – that can incline a person to the sorrow of acedia. Chapter Three presents an original definition of instrumental rationality, drawing from common understandings of instrumental reason as well as Max Weber's concept: instrumental rationality is a persistent disposition by which a person emphasizes their successful (e.g. effective, efficient, maximal) attainment (e.g. acquisition, possession, control, use) of intermediate goods and means that further one or more given ends – rather than attending to the choice and guidance of appropriate (e.g. good, moral, fulfilling) ends. The definition distinguishes instrumental rationality from the notion of a purely mental and procedural kind of instrumental reasoning. It therefore avoids the interminable philosophical debates over the nature, possibility, and roles of a specific type of reason described as instrumental, practical, or means-end. It also focuses us on a disposition that involves man's will along with the subsequent cognitive and behavioral tendencies. As demonstrated in the summary of comments about instrumental rationality by recent popes and related passages of the Bible, instrumental rationality has a strong and often negative influence on the paradigms and passions that underlie immoral behavior. Instrumental rationality is a tendency of the subject to focus on the efficiency or efficacy of the means such that it undermines or distorts the given end. With habituation, this self-deception may further devolve into the self-deception of acedia by which the intermediate means to man's excellence may be falsely apprehended to be in opposition with the divine good as participated by man.

In Chapter Four, we see that, in a variety of ways, AI technology induces instrumental rationality as a disposition. This manifests in such a way that the disposition in persons exposed to the technology may easily sway them toward the self-deception of acedia. We also see that instrumental rationality is, in a sense, structurally integral to the very nature of AI and its applications.

Several manifestations of instrumental rationality and its path to acedia are described in Chapter Five. The examples are not proofs, not even in an inductive sense, but they serve as various case studies that illustrate the actual effects of AI and its logical connection to instrumental rationality, with further deductive arguments for the possibility of motivating acedia. The extent of the examples, their deleterious effects, and the potential for AI to instigate acedia should, however, be convincing. The reader likely will imagine many other damaging effects of AI that are currently or likely to be experienced by our society, such as immersion in digitally generated "virtual" worlds, the privacy intrusions associated with brain-machine interfaces that enable two-way communication between personal thoughts and computer machine operations, enhanced distortions of identities and relationships through internet-based social media, the deterioration of habitual skills in writing and reasoning, etc.

This work points the way toward an understanding of modern technology as a structural, motivating, and contextual factor in the prevalence of sin in our society. In much of today's theological discussions about AI and modern technology, there is an emphasis on technology as a relationship of power, as Romano Guardini theorized.[478] While there is certainly reason to evaluate technology as manifesting relationships between the engineers, developers, owners, users, and other human objects of technological purposes – and analysis of relationships of power, coercion, and ideology are crucial to understanding how we might help the most disadvantaged groups in our society, from human embryos as test subjects to workers displaced by reckless technological and economic "progress" – theologians' efforts are also furthered by attention to the structural demands and sanctions (or incentives) that characterize specific technologies and how they can motivate dispositions to sin or to virtue. Contextual analysis of the effects of these technologies also takes us away from overly general attributions of evil or goodness to technology and to specific technologies; with contextual specificity, we can see that certain technologies are by no means morally neutral but powerfully influence moral or cognitive dispositions in man, who is always free in the final analysis to apply his reason to his consent or rejection of his inclinations as long as he is not afflicted by ignorance of the conditions and their influences.

The Church therefore has a great opportunity to educate Christians and others of the complex motivational relationship between AI and sin. This pastoral approach goes beyond merely countering a troublesome ideology of technological exceptionalism and progress; it is not associated with the myth of a false, hegemonic ideology propagated by deterministic philosophies that imagine the world as mired in conflict over ubiquitous power structures focused on material or cultural interests. It is instead an analysis of the structural and logical tendencies that characterize AI technologies, evaluation of the contextual ends of the

---

[478] Romano Guardini, *Power And Responsibility: A Course Of Action For The New Age* (Washington, DC: Regnery, 1961); and *The End of the Modern World* (Washington, DC: Regnery, 1961).

persons and corporate entities involved, and application of such under-standing to the personal experiences of individuals, their efforts to de-velop the habits of virtue, and most importantly their loving relation-ship with God through Christ.

Countering AI-Induced Acedia

A central theme of this book is that a society and culture saturated with AI technology runs the great risk of widespread acedia. Consistent use and experience of the technology do not fully determine Christians' behavior, but they structure, incentivize, and restrict it in a decidedly negative way. There is an ideological and habitual force to AI that mo-tivates sin, especially – and perhaps most consequentially – the sin of acedia.

There seems to be no escaping the fact that reducing and altering our experience of AI technology is crucial to building a life of virtue in the 21st century. This is an extremely difficult effort when AI is so per-vasive and also alluring in its nearly endless applications and frenetic pace of development. It is likely that many people in our society are un-concerned about either Christian virtue or limiting access to the won-derful applications, entertainment, and opportunities for societal trans-formation that AI will increasingly offer. It is likely that many in our secular society, yearning for some kind of personal relationship with their world that brings security, identity, and peace, will be all too happy to embrace anthropomorphic fantasies about AI and the machines it animates, building a social world oriented around interactions with AI robots and chatbots. The siren call of AI-powered creation of virtual "universes" will draw many toward socially, psychologically, and spirit-ually distorted manipulations of personal identities and the experience of reality itself. Will virtuous Christians find themselves enclosed and bewildered within this hyper-technological storm? Or, in our retreat from such a dangerous culture, will we find ourselves increasingly isolated and unable to interact with – or evangelize to – the AI enthusiasts and consumers all around us?

To wrestle with this dilemma, we need first of all to develop very clear principles, drawn from Christian wisdom and faith, that guide the mass of Christian faithful in understanding just what, how, and why AI has an influence on our spiritual destiny. The 2025 document *Antiqua et Nova* "On the Relationship between Artificial Intelligence and Human Intelligence" from the Dicastery for the Doctrine of the Faith and the Dicastery for Culture and Education is a very good start.[479] It offers a theological and thorough analysis of AI, especially of "social" interactions with AI systems, that an educated Christian should be able to absorb and reflect on. It also presents a thorough evaluation of the notion of artificial "intelligence" and contrasts it with human intelligence. Unfortunately, the document seems to emphasize problems with the bad use of AI, as if AI technology itself has a passive role in encouraging those bad uses, and understates the structural and ideological influence of AI on the motivations of users, producers, and developers. We need a comprehensive, more broadly referenced, technically savvy, and philosophically, theologically, and Biblically wise statement that the public can depend on as a light in the darkness, and we need it urgently.

We also need an intense conversation among Christians, and between Christians and the broader culture, about the limits that ought to be placed on the integration of AI technology in our lives. This might include moral justifications (or prohibitions) for policies related to protection of youth from damaging content and anthropomorphic interactions; consumer rights such as the right to repair devices and guarantees

---

[479] Dicastery for the Doctrine of Faith and Dicastery for Culture and Education, *Antiqua et Nova.*

of the quality, consistency, and persistence of certain critical technologies like brain-machine interfaces; weapons development and use; environmental degradation; access to and design of "virtual" worlds, games, and simulations; distribution of technology and its benefits to poorer nations and populations; surveillance by corporations and governments; etc. There is certainly some perceptive writing and thought available in the Christian community about such topics, but it seems that the Christian voice is a markedly weak, confused, and divided participant in societal dialogue (with the exception of Pope Francis' leadership in bringing some of the dangers of AI to the world's attention). Having a political and policymaking impact on the nature of the overall technological environment in which we live, worship, and raise our children is tremendously important when AI has such powerful ideological and moral influences. It will be better to shape the technological environment now rather than deal with the aftermath of purely secular and commercial leadership on such issues.

Given the likelihood of an extensive, if not ubiquitous, presence of AI in our daily lives, we will also need understanding and skills to counter the vice of instrumental rationality. There seem to be three general ways to deal with such a vice: 1) to be vigilant about its presence and then actively resist it by one's will; 2) to engage in contrary (virtuous) behavior; and 3) to pray for and cultivate the cardinal virtues of prudence, justice, fortitude, and temperance, since all virtues are built upon these "hinge" habits.

Vigilance – that is, watchful and intentional self-control – will be enhanced or at least enabled by education and further scholarship related to the vice of instrumental rationality. We should analyze other technologies and develop a clear set of definitions that help us to identify when and how the vice is motivated in different technological contexts. It will also help for Christian theologians and philosophers to build an understanding of how instrumental rationality has come to be such a commonly accepted feature of our culture, and this requires drawing on analyses of instrumental rationality in non-technological as well as technical facets of life. In particular, our theological analysis of the vice can inform as well as interpret sociological and economic evaluations of consumer behavior, professional standards, moral ideologies like contemporary utilitarianism, capitalist structures and influences on individual actors, etc. Much social scientific work has already been done, for example, by 20th century writers in the Frankfurt School and by their followers to highlight the role of instrumental rationality as a perversion of reason in late capitalism and fascism; this is an area of thought that requires sifting through biased political and ideological leanings as well as a virulently secular mindset, yet one that offers much intellectual wealth to a faith-based research agenda.[480] Our emphasis must be on instrumental rationality as a moral vice, not as a merely intellectual distortion or ideology that is confined to certain historical and societal developments.

Within theological discourse, we should also learn to recognize and refer to instrumental rationality as a vice, particularly when evaluating the role of advanced technology in society. The term has the benefit of merging ideological, moral, and technical considerations in a unified experience of habitual inclination to sin. I believe that Pope Francis' "technocratic paradigm" bears many of the same features as instrumental rationality, yet it remains ambiguous in its nature as either an intellectual ideology, individual habit, political force, or structural component of technology. The vice of instrumental rationality is, in many ways, an intellectual vice, yet it is primarily a habit of behavior with strong moral implications.

---

[480] See, for example, Max Horkheimer and Theodor W. Adorno, *Dialectic of Enlightenment* (Stanford, California: Stanford University Press, 2002).

Aside from learning to be vigilant, if we are to successfully engage in virtuous behavior that is contrary to instrumental rationality, thereby defeating the habit, we'll need to first identify what virtue is contrary. It seems that prudence is a good candidate. This virtue is "right reason about things to be done," disposing the virtuous person to choosing good actions in contingent situations.[481] It is not merely an intellectual trait, for it concerns the ends of action which are good, and this is a moral concern that involves the desires of the will.[482] In operation, prudence not only draws upon the various kinds of logic associated with rational thought, but also on the human person's natural knowledge of the highest moral principles, called synderesis; "In this way, the law of our understanding is related to prudence as an indemonstrable principle is to a demonstration."[483] Like instrumental rationality, then, prudence is a semi-intellectual virtue. As a cardinal virtue, prudence is crucial for the formation and exercise of other virtuous behaviors; instrumental rationality similarly engenders many other vices. Prudence, however, refers directly to the right and good exercise of a holistic reason, while instrumental rationality is narrow, diverted, and ultimately foolish. Instrumental rationality is decidedly opposed to the guidance of natural knowledge of moral principles within the process of selecting appropriate means for a given end. Prudence cannot be attained when succumbing to the vice of instrumental rationality because the rectitude of the inclinations of the will is not subjected to any truth standard. Practical ends become contingent, subjective, and morally relative.

If we are to counter the vice of instrumental rationality, we should encourage Christians to engage in activities that develop the habit and related skills (both reasoning and moral discernment) of prudence.

---

[481] Aquinas, *Summa Theologica* I-II, q.65, a.1, ad 4.

[482] Aquinas, *Summa Theologica* I-II, q.65, a.1; q.58, a.5, ad 1.

[483] Thomas Aquinas, *Quaestiones Disputatae de Veritate*, 5, 1, ad 6.

Practical reasoning skills can only come with real or hypothetical exposure to a variety of contingent situations, informed by the guidance of
wise counselors and teachers. This is the Aristotelian *phronesis*, the ac-
cumulation of practical wisdom through the development of positive
habits. The habitual edifice on which wisdom is built can, however, de-
teriorate with a lack of practice, as St. Thomas Aquinas worried.[484]

Much has been written and practiced in regard to youth education
for this purpose; I won't repeat this wisdom here. Saturation of AI technology in the classroom as well as the broader society will make it more
difficult to engage persons interpersonally, and this technology-medi-
ated contact will limit the relational experience, conversation, debate,
and role modeling that have traditionally characterized virtues education. For this reason, perhaps more than others, educators and evange-
lists alike should thoughtfully resist the desire to cloak themselves in
the spirit of the times by over-utilizing AI-enabled devices in teaching
and communication. This is not about blindly insisting on tradition, but
about retaining the full range of interpersonal and situational conditions that enable persons to encounter moral dilemmas with wise guidance and to both discern and thereby learn good actions.

Particularly in the section regarding "mediocrity" in Chapter 5, we
saw a number of ways that AI technology actually replaces the implementation of reasoning skills and can thereby cause such skills to atro-
phy. Christians should educate the public on such dangers as well as the
timeless value of the virtue of prudence itself. An assertive program of
restoring Aristotelian ethics in college philosophy programs and even
secondary schools would go a long way toward mitigating the cultural
effects of AI. Such a formation effort might lack the crucial element of
natural knowledge of the good that is essential for Christian prudence,

---

[484] Aquinas, *Summa Theologica* I–II, q.53, a.3.

but it would provide our evangelization efforts with a secure philosophical foundation in common sense. More broadly, Christians should make explicit promotion of the virtue of prudence a central element of the unified, public defense against hyper-technological extremes in our society; the virtue is strongly represented in Western tradition and its worth still resonates, I believe, among even the most assertive advocates of secular ethics.

Despite any such efforts at countering the vice of instrumental rationality, the inescapable presence of AI technology in our daily lives, social relations, and culture will quite likely encourage the sorrow and sin of acedia. Fortunately, there has been much written about the remedies for acedia. Evagrius Ponticus recommends perseverance, or hypomonè, whether in prayer, manual labor, or other activities. The point is to resist the restlessness and anxiety of acedia and learn both the discipline and the peace of remaining in one "place" (e.g., location, status, condition, etc.) over an extended period of time. Such perseverance is not simply a self-denial or punishment, but it is a very special opportunity to overcome self-love and truly advance in spiritual health.[485] Evagrius writes: "If the spirit of ακεδια comes over you, do not leave your dwelling (cf. Eccles. 10: 4) or avoid a worthwhile contest at an opportune moment, for in the same way that one might polish silver, so will your heart be made to shine."[486] Perseverance is not an experience of unbearable hardship, but one of openness to a lasting relationship with God, as Jesus taught: "Come to me, all you who labor and are burdened, and I will give you rest. Take my yoke upon you and learn from me, for I am meek and humble of heart; and you will find rest for ourselves. For my yoke is easy, and my burden light" (Matthew 11: 28-30).

---

[485] Nault, "Acedia: Enemy of Spiritual Joy," 239.
[486] Evagrius, *Praktikos* 28, 102.

In the context of an AI-induced acedia, another form of perseverance will involve a thoughtful and resolute – not stubborn and indis-
criminate – refusal to choose, use, encourage, or otherwise support the
hyper-technological saturation of daily life and social or governing
structures with AI technology. In healthcare, for example, where there
are tremendous opportunities to marshal the powers of AI for the reduction of suffering, we do not have to allow AI systems to take over the
predictive analytics of insurers, the administration of hospitals, or the
dictation and parsing of medical notes during patient visits. A virtuous
life with AI will be one of confident, peaceful, responses of "no" to the
eager and breathless adoption of every potential AI application.

Stability, in fact, seems to be the virtue most recommended in countering acedia, by both ancient and modern commentators. Even manual
labor, a favorite remedy for Cassian, is intended to develop zeal in the
monk's relationship to God and others in the community and courage
to stay in place rather than take the supposedly easy route of flight or
changing course.[487] Aside from attentive work, a person could just as
fruitfully choose true leisure in the effort to combat acedia, as described
by Josef Pieper:

> Because wholeness is what man strives for, the
> power to achieve leisure is one of the fundamental pow-
> ers of the human soul. Like the gift for contemplative
> absorption in the things that are, and like the capacity
> of the spirit to soar in festive celebration, the power to
> know leisure is the power to overstep the boundaries of
> the workaday world and reach out to superhuman life-

---

[487] Dennis Ockholm, "Staying Put to Get Somewhere."

giving existential forces that refresh and renew us before we turn back to our daily work.[488]

Pieper links his promotion of the remedy of celebratory leisure to
the command to honor the Sabbath, a command that St. Thomas Aquinas also highlights in relation to acedia.[489] For St. Thomas, the person
who succumbs to acedia violates the spirit and letter of the Sabbath
command, "for this precept, in so far as it is a moral precept, implicitly
commands the mind to rest in God: and sorrow of the mind about the
Divine good is contrary thereto."[490]

The Sabbath helps us to direct our rapt and sober attention to that
which is most important in our lives. This act of truly seeing both the
spiritual and physical reality of our world is a crucial antidote to the
distraction that characterizes acedia. Here, perseverance is needed not
only to maintain stability but to actively seek the truth through contemplation. Kyle Childress reminds us of the example of Mary Magdalene,
who had the great honor of seeing the resurrected Christ because she
persistently stood by his tomb and waited with mournful yet watchful
eyes.[491] Such attentiveness also underlies the "culture of care" that Pope
Francis earnestly calls for; our loving attention to the needs of our environment and of other persons draws us into a relationship of responsible, reverent stewardship for all creation.[492] It pulls us out of the self-
deception of acedia that sets up a false opposition between 1) our individual welfare and subjective consciousness and 2) the integral

---

[488] Josef Pieper, *Leisure: The Basis of Culture* (San Francisco: Ignatius Press, 2009), 50-51.

[489] Aquinas, *Summa Theologica* II-II, q.35, a.3, ad 1.

[490] Aquinas, *Summa Theologica* II-II, q.35, a.3, ad 1.

[491] Kyle Childress, "Sloth: Who Cares?" in Kruschwitz, *Acedia*, 73-6.

[492] Francis, *Laudato si'*, 228-32.

development of humanity and of the objective reality around us. Such attention is not purely outward-oriented, for it changes the seeker and the seer whose eyes, mind, and heart absorb that which is attended to. Christopher Blum and Joshua Hochschild explain:

> For this reason, the common English phrase "to pay attention" is misleading. We do not trade or exchange our attention, we give it. As with most gifts, this means we give something of ourselves in the attention, and we find that what we give attention to gives something of itself back. It is also not merely an investment of our thoughts, but also of our emotions and most especially of our will. Attention is, accordingly, an investment of our very self.[493]

In regard to our relationship with technology, Albert Borgmann helpfully encourages us to develop and fix our attention on "focal practices." Rather than passively allow AI-enabled devices and applications to mesmerize us and excite our consumer sensibilities, we can find alternative things to do that we love and that enhance the more important values in our lives. "Focal things and our engagement with them orient us and center us in time and space in ways that technological devices do not. A focal thing is not at the mercy of how you feel at the moment, whether the time is convenient or whatever; you commit yourself to it come hell or high water."[494] An example of a focal thing might be a

---

[493] Christopher O. Blum and Joshua P. Hochschild, *A Mind at Peace: Reclaiming an Ordered Soul in the Age of Distraction* (Manchester, New Hampshire: Sophia Institute Press, 2017).

[494] Albert Borgmann, in "Albert Borgmann on Taming Technology: An Interview," *The Christian Century* (August 23, 2003), 22-5.

roaring fireplace in one's home and the focal practice might be sharing stories among the family or playing a particular game. The temptation of technology is its low "threshold" or minimal barriers to gaining immediate, but low-value satisfaction, while focal practices may involve high thresholds but also high value and rewards.[495] Ultimately, attention to focal practices requires devotion, initial willpower, habituation, and clear awareness of the high value.

The Eucharist is, of course, a focal practice for Catholics, and it reminds us that we have need of spiritual nourishment; just as much as we are called to pay attention to living our virtuous lives in this world, we must also heed Jesus' teaching that "You are not of the world" (Jn 15:19).[496] Jean-Charles Nault sees the Incarnation and our attention to it as perhaps the best remedy for acedia. According to Nault, "The Eucharist is what gives temporality its ultimate meaning, since it takes up the past, the present, and the future: love never passes away (1 Cor 13:8)."[497] St. Thomas Aquinas also indicates that the Incarnation is the act and the truth that directly opposes the sorrow of acedia and its self-deception of an unbridgeable divide between the human being and their God:

> Countering man's despair at the enormity of a vocation that he feels unable to achieve, the Incarnation of Christ offers a new principle of action that rescues man from the *taedium operandi* and allows him to open his heart once again to the gift of divine friendship. Christ,

---

[495] Albert Borgmann, in "Albert Borgmann on Taming Technology: An Interview."

[496] Nault, "Acedia: Enemy of Spiritual Joy," 257-8.

[497] Jean-Charles Nault, *The Noonday Devil: Acedia, the Unnamed Evil of Our Times*, transl. Michael J. Miller (San Francisco: Ignatius Press, 2013), 142.

both true God and true man, achieves within himself, in a singular and unique way, the union between Creator and creature that God desired and to which man is called, if he agrees to open himself to the gift of divine friendship.[498]

Rather than striving futilely to acquire and earn God's love through human, instrumentally oriented effort, the person learns through the Incarnation to trust in God's friendship and to release their attention – to open themselves to it. This is a matter of trusting in one's own inherent, created dignity as much as it is hoping for fulfillment of some transaction. As Pieper describes it, a person afflicted with acedia "does not want to be what God wants him to be, and that means that he does not want to be what he really, and in the ultimate sense, is."[499] Such a false perspective is impossible if bathed in the light of the Incarnation and Resurrection of Christ.

Another way to learn and reinforce this sense of personal dignity is to engage in authentic relationships with others, unmediated by restrictive and instrumentally-oriented technologies.[500] Along with authenticity, this communication is characterized by hospitality and forgiveness.[501] This is why Pope Francis calls us to a culture of encounter with others: "The Gospel tells us constantly to run the risk of a face-to-face encounter with others, with their physical presence which challenges us, with their pain and their pleas, with their joy which infects us

---

[498] Nault, "Acedia: Enemy of Spiritual Joy," 246-7.

[499] Pieper, Leisure, 28; Amy Freeman, "Remedies to Acedia in the Rhythm of Daily Life," in *Acedia*, edited by Kruschwitz, 40.

[500] Ockholm, "Staying Put to Get Somewhere."

[501] Jonathan Wilson-Hartgrove, *The Wisdom of Stability: Rooting Faith in a Mobile Culture* (Brewster, Massachusetts: Paraclete Press, 2010), 18.

in our close and continuous interaction."[502] Such authentic communication and community is hardly consonant with an AI-dominated world. The Church can serve as a cultural and communal alternative that offers the kind of loving, in-person relationships that many will crave in an AI-dominated society. This kind of communication enhances social interactions and solidarity, but even more so it brings the individual participants into a greater holiness:

> It is not simply a matter of making machines appear more human, but of awakening humanity from the slumber induced by the illusion of omnipotence, based on the belief that we are completely autonomous and self-referential subjects, detached from all social bonds and forgetful of our status as creatures.[503]

While my emphasis in this concluding chapter is on virtuously circumscribing the role and influence of AI in our daily lives, it is possible to more fruitfully integrate the technology. I agree with Nadia Delicata that, particularly in the context of advanced technology, we should modify the traditional contrast between *phronesis* as a virtuous process of habituating a person to act and orient themselves toward what is good – that is, the formation of wisdom – and *techne* as mere craftsmanship that is oriented only to development of the product.[504] We should bear in mind that craftsmanship also involves habituation to the

---

[502] Francis, *Evangelii Gaudium* "On the Proclamation of the Gospel in Today's World" (November 24, 2013), 88.

[503] Francis, Message for the 58th World Day of Social Communications "Artificial Intelligence and the Wisdom of the Heart: Towards a Fully Human Communication" (January 24, 2024).

[504] Nadia Delicata, "Natural Law in a Digital Age," *Journal of Moral Theology* 4, no.1 (2015).

techniques and standards of the craft, and it usually calls for attention to the values, cultural meanings, and full lived experiences of those others who will engage with the product. Creating our products is more or less a process of self-transformation (I would add that it is also a socially negotiated process of cultural innovation).

In the AI environment, the design, development, marketing, and shared use of AI-enabled products very often requires all those involved to immerse themselves in the cultural and moral characteristics of other actors. For example, the content of databases that LLM models rely on is reflective of societal values and biases; the developers of AI-enabled, augmented reality eyeglasses must navigate the expectations of the consumers about privacy protection; virtual reality games and social simulations will be populated and negotiated by real persons who make moral choices about everything from their behavior to their communications and apparent identities; and corporate executives will continue to bear responsibility for the many thousands of low-paid workers who handle data labeling and re-labeling of graphic or violent content in LLM databases and therefore experience mental health issues from such exposure. The entire milieu of AI-enabled structures, systems, and applications is a network of persons – craftspersons, we might say – who influence and are deeply influenced by their involvement. Although Delicata does not directly address the AI-driven, hyper-technological context, her conclusion rings true: "Thus, the reasonability of techne is not simply a matter of method and efficient causality. It must also be considered from the point-of-view of its finality, that is, through the inherent intelligibility and beauty of what it makes—ultimately the re-creation of the human and the cosmos itself."[505]

---

[505] Delicata, "Natural Law in a Digital Age," 20.

At this time, even the oppressive and opaque systems of AI rely for their implementation on the willing acceptance, guidance, and selective adoption of applications by a wary, even fearful, yet dutifully consuming public. We have the opportunity, then, to transform or at least minimize the instrumentally rational character of AI technology. This occurs in every interaction we have with the AI context by which we make a moral choice – that is, we pay close attention to the systems, devices, and other human actors involved in their use and development; we discern and orient ourselves to the good; we act with prudence; and we form habits of choice and behavior that reflect that orientation to the good. In doing so, we craft and re-craft our technological systems and the devices and applications that they support. As we bring prudence to our interactions with advanced technologies, the AI systems begin to appear less like neural nets or structures of pure information and more like structured social environments in which the free, moral actions of persons form nodes and layers of morally relevant values. We can either sorrowfully resign in the face of the ruthless instrumental rationality of technological systems and machines, superficially enjoying our relatively submissive existence as mere consumers of information, applications, and corporate governance, or we can insist on a counter-cultural orientation to the good and the true, forcing the peddlers of AI systems to design greater freedom of conscience by their users as well as more democratic selection of ends for AI applications. At no point will AI systems develop either the general wisdom or the moral judgment to do this autonomously; this requires intuition of moral principles, the natural knowledge of practical goodness derived through synderesis. Christians must actively assume the mantle of stewardship of both God's creation and our own if we want to defend the broad sphere of moral agency through which we grow in virtue, holiness, and friendship with our Creator.

Did you enjoy this book?

Please share a rating and review online!

Bibliography

Adam, Hammaad, Aparna Balagopalan, Emily Alsentzer, Fotini Christia, and Marzyeh Ghassemi. "Mitigating the Impact of Biased Artificial Intelligence in Emergency Decision-Making." *Communications Medicine* 2 (2022): 149. https://doi.org/10.1038/s43856-022-00214-4.

A.I. Research Group for the Centre for Digital Culture of the Dicastery for Culture and Education of the Holy See. *Encountering Artificial Intelligence: Ethical and Anthropological Investigations.* Edited by Matthew J. Gaudet, Noreen Herzfeld, Paul Scherz, and Jordan J. Wales. Eugene, Oregon: Pickwick Publications, 2024.

Aijian, J. L. "Fleeing the Stadium: Recovering the Conceptual Unity of Evagrius' Acedia." *Heythrop Journal* 62, no.1: 7-20.

Akter, Shahriar, Yogesh K. Dwivedi, Shahriar Sajib, Kumar Biswas, Ruwan J. Bandara, and Katina Michael. "Algorithmic Bias in Machine Learning-Based Marketing Models." *Journal of Business Research* 144 (2022): 201–216. https://doi.org/10.1016/j.jbusres.2022.01.083.

Akter, Shahriar, Grace McCarthy, Shahriar Sajib, Katina Michael, Yogesh K. Dwivedi, John D'Ambra, and K. N. Shen. "Algorithmic Bias in Data-Driven Innovation in the Age of AI." *International Journal of Information Management* 60 (2021): 102387. https://doi.org/10.1016/j.ijinfomgt.2021.102387.

American Psychiatric Association. "American Adults Express Increasing Anxious-ness in Annual Poll; Stress and Sleep are Key Factors Impacting Mental Health" (May 1, 2024). https://www.psychiatry.org/news-room/news-releases/annual-poll-adults-express-increasing-anxiousness.

Ammari, T., J. Kaye, J. Y. Tsai, and F. Bentley. "Music, Search, and IoT: How People (eRally) Use Voice Assistants." *ACM Transactions on Computer-Human Interaction*, 26, no.3 (2019): 1–28.

Aquinas, Thomas. *Commentary on St. Paul's Letter to the Galatians.* Translated by F. R. Larcher. Albany: Magi Books, 1966. https://isidore.co/aquinas/english/SSGalatians.htm.

Aquinas, Thomas. *Questiones Disputatae de Malo.* Translated by Richard Regan. Oxford: Oxford University Press, 2003. https://isidore.co/misc/Res%20pro

%20Deo/Logic/Logic%20Course%20Material/Aquinas%20Texts/Aqui-nas%20(Newer%20PDFs)/Quaestiones%20Disputatae%20de%20Malo%20ENGLISH.pdf.

Aquinas, Thomas. *Questiones Disputatae de Veritate.* In *The Collected Works of Thomas Aquinas.* Hastings: Delphi Classics, 2020.

Aquinas, Thomas. *Summa Contra Gentiles.* Translated by Charles J. O'Neil. Garden City, New York: Image Books, 1957.

Aquinas, Thomas. *Summa Theologiae.* Translated by the Fathers of the English Dominican Province. New Advent, 2017. https://www.newadvent.org/summa/.

Araujo, Theo. "Living Up to the Chatbot Hype: The Influence of Anthropomorphic Design Cues and Communicative Agency Framing on Conversational Agent and Company Perceptions." *Computers in Human Behavior* 85 (2018): 183–189. https://doi.org/10.1016/j.chb.2018.03.051.

Arias, Cristian Augusto Gonzalez. "ChatGPT's Artificial Empathy Is a Language Trick. Here's How It Works." *TechXplore* (November 28, 2024). https://techxplore.com/news/2024-11-chatgpt-artificial-empathy-language.html.

Ashton, Hal and Matija Franklin. "The Problem of Behaviour and Preference Manipulation in AI Systems." *CEUR Workshop Proceedings* (2022): 3087. https://ceur-ws.org/Vol-3087/paper_28.pdf.

Augustine, *On Christian Doctrine* I. Translated by James Shaw. In *Nicene and Post-Nicene Fathers*, First Series, Vol. 2. Edited by Philip Schaff. Buffalo, NY: Christian Literature Publishing Co., 1887. Revised and edited for New Advent by Kevin Knight. http://www.newadvent.org/fathers/1202.htm.

Augustine. *The City of God.* Translated by Marcus Dods. New York: The Modern Library, 2000.

Axelrod, Robert. *The Complexity of Cooperation: Agent-Based Models of Competition and Collaboration.* Princeton: Princeton University Press, 1997.

Balthasar, Hans Urs von. *Theo-Drama: Theological Dramatic Theory: The Action*, vol. 4. Translated by Graham Harrison. San Francisco: Ignatius Press, 1994.

Beede, Emma, Elizabeth Baylor, Fred Hersch, Anna Iurchenko, Lauren Wilcox, Paisan Ruamviboonsuk, Laura M. Vardoulakis, et al. "A Human-Centered Evaluation of a Deep Learning System Deployed in Clinics for the

Detection of Diabetic Retinopathy." *CHI '20: Proceedings of the 2020 CHI Conference on Human Factors in Computing Systems* (2020): 1-12. https://doi.org/10.1145/3313831.3376718.

Bender, Emily M., Timnit Gebru, Angelina McMillan-Major, and Shmargaret Shmitchell. "On the Dangers of Stochastic Parrots: Can Language Models Be Too Big?" *Proceedings of the 2021 ACM Conference on Fairness, Accountability, and Transparency* (2021): 610–623. https://doi.org/10.1145/3442188.3445922.

Benedict XVI. Address to Participants in a Congress on "Digital Witnesses: Faces and Languages in the Cross-Media Age" (April 24, 2010). http://www.vatican.va/holy_father/benedict_xvi/speeches/2010/april/documents/hf_ben-xvi_spe_20100424_testimoni-digitali_en.html

Benedict XVI. Address on the Occasion of Christmas Greetings to the Roman Curia (December 20, 2010). https://www.vatican.va/content/benedict-xvi/en/speeches/2010/december/documents/hf_ben-xvi_spe_20101220_curia-auguri.html.

Benedict XVI. Address to Participants in a Congress on "Digital Witnesses, Faces and Languages in the Cross-Media Age" (April 24, 2010). http://www.vatican.va/holy_father/benedict_xvi/speeches/2010/april/documents/hf_ben-xvi_spe_20100424_testimoni-digitali_en.html.

Benedict XVI. Address to Participants in the Plenary Meeting of the Pontifical Council *Cor Unum* (January 19, 2013). https://www.vatican.va/content/benedict-xvi/en/speeches/2013/january/documents/hf_ben-xvi_spe_20130119_pc-corunum.html.

Benedict XVI. *Caritas in veritate* "On Integral Human Development in Charity and Truth." June 29, 2009.

Benedict XVI. *Deus caritas est* "On Christian Love." December 25, 2005.

Benedict XVI. "Faith, Reason and the University: Memories and Reflections," lecture (September 12, 2006). https://www.vatican.va/content/benedict-xvi/en/speeches/2006/september/documents/hf_ben-xvi_spe_20060912_university-regensburg.html.

Benedict XVI. *Spe salvi* "On Christian Hope." November 30, 2007.

Bertrand, J. and L. Weill. "Do Algorithms Discriminate against African Americans in Lending?" *Economic Modeling*, 104 (2021): 105619. https://doi.org/10.1016/j.econmod. 2021.105619.

Bjerring, Jens Christian and Jacob Busch. "Artificial Intelligence and Identity: The Rise of the Statistical Individual." *AI and Society* (2024). https://doi.org/ 10.1007/s00146-024-01877-4.

Blau, Wolfgang, et al. "Protecting Scientific Integrity in an Age of Generative AI." *PNAS* 121, no.22 (2024): e2407886121. https://doi.org/10.1073/pnas.2407886121.

Bless, H. and M. Wänke. "Can the Same Information Be Typical and Atypical? How Perceived Typicality Moderates Assimilation and Contrast in Evaluative Judgments." *Personality and Social Psychology Bulletin*, 26, no.3 (2000): 306–314.

Blum, Christopher O. and Joshua P. Hochschild. *A Mind at Peace: Reclaiming an Ordered Soul in the Age of Distraction*. Manchester, New Hampshire: Sophia Institute Press, 2017.

Bogost, Ian. "The Cathedral of Computation." *The Atlantic* (January 15, 2015). https://www.theatlantic.com/technology/archive/2015/01/the-cathedral-of-computation/384300/.

Booth, Robert. "AI Could Cause 'Social Ruptures' between People Who Disagree on Its Sentience." *The Guardian* (November 17, 2024). https://www.theguardian.com/technology/2024/nov/17/ai-could-cause-social-ruptures-between-people-who-disagree-on-its-sentience.

Borgmann, Albert. In "Albert Borgmann on Taming Technology: An Interview." *The Christian Century* (August 23, 2003): 22-5.

Boyer, Pascal. "What Makes Anthropomorphism Natural: Intuitive Ontology and Cultural Representations," *The Journal of the Royal Anthropological Institute* 2, no.1 (1996): 83. https://doi.org/10.2307/3034634.

Brake, Josh. "Fake Personableness." The Absent-Minded Professor blog (November 19, 2024). https://joshbrake.substack.com/p/fake-personableness.

Brandtzæg, Petter Bae, Marita Skjuve, Kim Kristoffer Dysthe, and Asbjørn Følstad. "When the Social Becomes Non-Human: Young People's Perception of Social Support in Chatbots." *CHI '21: Proceedings of the 2021 CHI Conference on Human Factors in Computing Systems*, no. 257 (2021): 1-13. https://doi.org/10.1145/3411764.3445318.

Brendel, Alfred Benedikt, Fabian Hildebrandt, Alan R. Dennis, and Johannes Riquel. "The Paradoxical Role of Humanness in Aggression Toward Conversational Agents." *Journal of Management Information Systems* 40, no.3

(2023): 883-913. https://doi.org/10.1080/07421222.2023.2229127.

Bromwich, Jonah Engel. "Why Do We Hurt Robots?" *New York Times* (January 19, 2019). https://www.nytimes.com/2019/01/19/style/why-do-people-hurt-robots.html.

Brummelman, Eddie, Stefanie A. Nelemans, Sander Thomaes, and Bram Orobio de Castro. "When Parents' Praise Inflates, Children's Self-Esteem Deflates." *Child Development* 88, no.6 (2017): 1799-1809. https://doi.org/10.1111/cdev.12936.

Brynjolfsson, Erik, Danielle Li, and Lindsey R. Raymond. *When and How Artificial Intelligence Augments Employee Creativity*. Cambridge, Massachusetts: National Bureau of Economic Research, 2023. DOI 10.3386/w31161. https://www.nber.org/papers/w31161.

Bunge, Gabriel. *Despondency: The Spiritual Teaching of Evagrius of Pontus*. Translated by Anthony P. Gythiel. Yonkers, New York: St. Vladimir's Seminary Press, 2011.

Burgess, Matt. "'AI Girlfriends' Are a Privacy Nightmare." *Wired* (February 14, 2024). https://www.wired.com/story/ai-girlfriends-privacy-nightmare/.

Caltrider, Jen, Misha Rykov and Zoë MacDonald. "Happy Valentine's Day! Romantic AI Chatbots Don't Have Your Privacy at Heart." Mozilla Foundation (February 14, 2024). https://foundation.mozilla.org/en/privacynotincluded/articles/happy-valentines-day-romantic-ai-chatbots-dont-have-your-privacy-at-heart.

Carroll, Micah, Rohin Shah, Mark K. Ho, Thomas L. Griffiths, Sanjit A. Seshia, Pieter Abbeel, and Anca Dragan. "On the Utility of Learning about Humans for Human-AI Coordination." *NIPS'19: Proceedings of the 33rd International Conference on Neural Information Processing Systems* (2019) 465: 5174–5185. https://dl.acm.org/doi/10.5555/3454287.3454752.

Cassian, John. *Conferences*. Translated by C.S. Gibson. In *Nicene and Post-Nicene Fathers*, Second Series, Vol. 11. Edited by Philip Schaff and Henry Wace. Buffalo, NY: Christian Literature Publishing Co., 1894. Revised and edited for New Advent by Kevin Knight. http://www.newadvent.org/fathers/3508.htm.

Cassian, John. *Institutes*. Translated by C.S. Gibson. Edited by Philip Schaff and Henry Wace. Revised and edited for New Advent by Kevin Knight. In *Nicene and Post-Nicene Fathers*, Second Series, Vol. 11. Buffalo, New York:

Christian Literature Publishing Co., 1894. http://www.newadvent.org/fathers/3507.htm.

Catholic Church. *Catechism of the Catholic Church*. Rome: Libreria Editrice Vaticana, 1997.

Celdir, Musa Eren, Soo-Haeng Cho, and Elina H. Hwang. "Popularity Bias in Online Dating Platforms: Theory and Empirical Evidence." *Manufacturing & Service Operations Management* 26, no.2 (2023): 537-553. https://doi.org/10.1287/msom.2022.0132.

Chan, Stephanie C.Y., Adam Santoro, Andrew K. Lampinen, and Jane X. Wang. "Data Distributional Properties Drive Emergent In-Context Learning in Transformers," in *Advances in Neural Information Processing Systems 35 (NeurIPS 2022)*. Edited by S. Koyejo, S. Mohamed, A. Agarwal, D. Belgrave, K. Cho, and A. Oh. 2022. https://proceedings.neurips.cc/paper_files/paper/2022/file/77c6ccacfd9962e2307fc64680fc5ace-Paper-Conference.pdf?utm_source=substack&utm_medium=email.

Chapekis, Athena, Samuel Bestvater, Emma Remy, and Gonzalo Rivero. "When Online Content Disappears." Pew Research Center (May 17, 2024). https://www.pewresearch.org/data-labs/2024/05/17/when-online-content-disappears

Cheong, I., Y. E. Huh, and S. Puntoni. "Consumers' Lay Beliefs about AI Evaluation of Interpersonal Skills." In *Proceedings of the Association for Consumer Re-search Conference*. Edited by L. Chaplin, P. Raghubir, and K. Wilcox. Duluth, Minnesota: Association for Consumer Research, 2023.

Childress, Kyle. "Sloth: Who Cares?" In *Acedia*. Edited by Robert B. Kruschwitz. Waco, Texas: The Center for Christian Ethics at Baylor University, 2013. 73-76.

Choi, Hanbyul, Jonghwa Park, and Yoonhyuk Jung. "The Role of Privacy Fatigue in Online Privacy Behavior." *Computers in Human Behavior* 81 (2018): 42-51. https://doi.org/10.1016/j.chb.2017.12.001.

Choi, Jonathan H. and Daniel Schwarcz. "AI Assistance in Legal Analysis: An Empirical Study," *Journal of Legal Education* 73 (2024). https://dx.doi.org/10.2139/ssrn.4539836.

Congregation for the Doctrine of the Faith. *Instruction "Donum Vitae" on Respect for Human Life in Its Origin and on the Dignity of Procreation: Replies to Certain Questions of the Day* (February 22, 1987).

Conybeare, Will and Rachel Menitoff. "Vandals, Thieves Attacking L.A. Food Delivery Robots." KTLA News (August 8, 2023). https://ktla.com/news/local-news/food-delivery-robots-under-attack-from-vandals-thieves-local-businesses-starting-to-be-affected/.

Corbyn, Zoë. "AI Scientist Ray Kurzweil: 'We Are Going to Expand Intelligence a Millionfold by 2045.'" *The Guardian* (June 29, 2024). https://www.theguardian.com/technology/article/2024/jun/29/ray-kurzweil-google-ai-the-singularity-is-nearer.

Cornwell, James F. M. and E. Tory Higgins. "Sense of Personal Control Intensifies Moral Judgments of Others' Actions." *Frontiers in Psychology* 10 (2019): 2261. https://doi.org/10.3389/fpsyg.2019.02261.

Cotton, Debby R. E., Peter A. Cotton, and J. Reuben Shipway. "Chatting and Cheating: Ensuring Academic Integrity in the Era of ChatGPT." *Innovations in Education and Teaching International* 61, no.2 (2024): 228-239. https://doi.org/10.1080/14703297.2023.2190148.

Courtois, Cédric and Elisabeth Timmermans. "Cracking the Tinder Code: An Experience Sampling Approach to the Dynamics and Impact of Platform Governing Algorithms." *Journal of Computer-Mediated Communication* 23, no. 1 (2018): 1-16. https://doi.org/10.1093/jcmc/zmx001.

Cymek, Dietlind Helene, Anna Truckenbrodt, and Linda Onnasch. "Lean Back or Lean In? Exploring Social Loafing in Human–Robot Teams." *Frontiers in Robotics and AI* 10 (2023). https://doi.org/10.3389/frobt.2023.1249252.

Dahm, Brandon. "Correcting Acedia through Wonder and Gratitude." *Religions* 12, no.7 (2021): 458-72.

Daly, Robert W. "Before Depression: The Medieval Vice of Acedia." *Psychiatry* 70, no. 1 (2007): 30-51.

Danaher, J. "Robot Betrayal: A Guide to the Ethics of Robotic Deception." *Ethics of Information Technology* 22, no.2 (2020):117–128. https://doi. org/10. 1007/ s10676- 019- 09520-3.

Debatin, Bernhard, "Ethics, Privacy, and Self-Restraint in Social Networking." In *Privacy Online*. Edited by Sabine Trepte and Leonard Reinecke. Berlin: Springer, 2011. 47-60. https://doi.org/10.1007/978-3-642-21521-6_5.

Delicata, Nadia. "Natural Law in a Digital Age." *Journal of Moral Theology* 4, no. 1 (2015): 7-9.

Denning, Peter J. "Can Generative AI Bots Be Trusted?" *Communications of the*

ACM 66, no.6 (2023): 24–27. https://doi.org/10.1145/3592981.

DeRossett, Tommy, Donna J. LaVoie, and Destiny Brooks. "Religious Coping Amidst a Pandemic: Impact on COVID-19-Related Anxiety." *Journal of Religion and Health* 60 (2021): 3161-3176.

Descartes, René. *Philosophical Letters.* Translated by Anthony Kenny. Minneapolis: University of Minnesota, 1970.

DeYoung, Rebecca Konyndyk. "Aquinas on the Vice of Sloth: Three Interpretive Issues." *The Thomist* 75, no. 1 (2011): 43-64.

Dezfouli, Amir, Richard Nock, and Peter Dayan. "Adversarial Vulnerabilities of Human Decision-Making." *PNAS* 117, no.46 (2020): 29221-29228. https://doi.org/10.1073/pnas.2016921117.

Dicastery for the Doctrine of the Faith and Dicastery for Culture and Education. *Antiqua et Nova* "On the Relationship between Artificial Intelligence and Human Intelligence" (January 28, 2025).

Diemerling, Hannes, Leonie Stresemann, Tina Braun, and Timo von Oertzen. "Implementing Machine Learning Techniques for Continuous Emotion Prediction from Uniformly Segmented Voice Recordings." *Frontiers in Psychology* 15 (2024). https://doi.org/10.3389/fpsyg.2024.1300996.

Dillion, Danica, Niket Tandon, Yuling Gu, and Kurt Gray. "Can AI Language Models Replace Human Participants?" *Trends in Cognitive Sciences* 27, no.7 (2023): 597-600. https://doi.org/10.1016/j.tics.2023.04.008.

Dong, Junyi, Qingze Huo, and Silvia Ferrari. "A Holistic Approach for Role Inference and Action Anticipation in Human Teams." *ACM Transactions on Intelligent Systems and Technology* 13, no.6 (2022):, 1-24. https://doi.org/10.1145/3531230.

Du, Ying Roselyn. "Personalization, Echo Chambers, News Literacy, and Algorithmic Literacy: A Qualitative Study of AI-Powered News App Users." *Journal of Broadcasting & Electronic Media* 67, no.3 (2023): 246–273. http://dx.doi.org/10.1080/08838151.2023.2182787.

Dumiak, Michael. "One AI to Another: Is That Your Best Offer?" *IEEE Spectrum* (November 7, 2024). https://spectrum.ieee.org/ai-contracts.

Eisikovits, Nir and Dan Feldman. "AI and Phronesis." *Moral Philosophy and Politics* 9, no.2 (2022).

Ellis, Steven. "The Varieties of Instrumental Rationality." *The Southern Journal of Philosophy* 46 (2008): 199-220.

Ellul, Jacques. *The Technological Society*. Translated by John Wilkinson. New York: Random House, 1964.

Epley, Nicholas, Adam Waytz, and John T. Cacioppo. "On Seeing Human: A Three-Factor Theory of Anthropomorphism." *Psychological Review* 114, no.4 (2007): 864. https://doi.org/10.1037/0033-295X.114.4.864.

Epstein, Joshua M. and Robert Axtell. *Growing Artificial Societies: Social Science from the Bottom Up*. Brookings Institution Press, 1996.

Eriksson, Thommy. "Design Fiction Exploration of Romantic Interaction with Virtual Humans in Virtual Reality." *Journal of Future Robot Life* 3 (2022), no.1: 63–75. DOI 10.3233/FRL-210007.

Espósito, Filipe. "iOS 15.5 Beta Blocks 'Sensitive Locations' for Memories in Photos App." *9 to 5 Mac* (April 26, 2022). https://9to5mac.com/2022/04/26/ios-15-5-beta-blocks-sensitive-locations-for-memories-in-photos-app/.

Evagrius Ponticus, *Kephalaia Gnostika: A New Translation from the Unreformed Text from the Syriac*. Translated by Ilaria L. E. Ramelli. Atlanta: SBL Press, 2015.

Evagrius Ponticus. "On the Eight Thoughts." In *Evagrius of Pontus: The Greek Ascetic Corpus*. Translated by Robert E. Sinkewicz. Oxford: Oxford University Press, 2003. 66-90.

Evagrius Ponticus. "On the Vices Opposed to the Virtues." In *Evagrius of Pontus: The Greek Ascetic Corpus*. Translated by Robert E. Sinkewicz. Oxford: Oxford University Press, 2006.

Evagrius Ponticus. *Reflections*. In *Evagrius: The Greek Ascetic Corpus*. Translated by Robert E. Sinkewicz. Oxford: Oxford University Press, 2003. 210-216.

Evagrius Ponticus. *The Praktikos and Chapters on Prayer*. Translated by John Bamberger. Piscataway, New Jersey: Gorgias Press, 2009.

Evans, Paul S. "Creation, Progress, and Calling: Genesis 1-11 as Social Commentary," *McMaster Journal of Theology and Ministry* (2011): 67-100.

Eyssel, Friederike and Natalia Reich. "Loneliness Makes the Heart Grow Fonder (of Robots)—On the Effects of Loneliness on Psychological Anthropomorphism. *2013 8th ACM/IEEE International Conference on Human-Robot Interaction* (2013): 121–122. https://doi.org/10.1109/HRI.2013.6483531.

Farrell, Henry. "After Software Eats the World, What Comes Out the Other End?" Programmable Mutter blog (October 3, 2024).

https://www.programmablemutter.com/p/after-software-eats-the-world-what.

Feathers, Todd. "AI Could Be Making Scientists Less Creative." *Gizmodo* (December 15, 2024). https://gizmodo.com/ai-could-be-making-scientists-lesscreative-2000538342.

Feenberg, Andrew. *Between Reason and Experience: Essays in Technology and Modernity*. Cambridge, Massachusetts: The MIT Press, 2010.

Feng, Shangbin, Chan Young Park, Yuhan Liu, and Yulia Tsvetkov. "From Pre-training Data to Language Models to Downstream Tasks: Tracking the Trails of Political Biases Leading to Unfair NLP Models." *Proceedings of the 61st Annual Meeting of the Association for Computational Linguistics, Vol.1: Long Papers* (2023): 11737–11762. https://aclanthology.org/2023.acl-long.656.pdf.

Fleder, D. and K. Hosanagar. "Blockbuster Culture's Next Rise or Fall: The Impact of Recommender Systems on Sales Diversity." *Management Science*, 55, no.5 (2009): 697–712.

Foster, Brian. "The Former CEO of Google Warns of a Global Catastrophe Caused by AI in Five Years." *Glass Almanac* (November 12, 2024). https://glassalmanac.com/the-former-ceo-of-google-warns-of-a-global-catastrophe-caused-by-ai-in-five-years/.

Fourcade, Marion and Kieran Joseph Healey. *The Ordinal Society*. Cambridge, MA: Harvard University Press, 2024.

Fox, Craig R., Michael Goedde-Menke, and David Tannenbaum. "Ambiguity Aversion and Epistemic Uncertainty" (2021). http://dx.doi.org/10.2139/ssrn.3922716.

Francis. Address to the Participants in the Seminar "The Common Good in the Digital Age." September 27, 2019. https://www.vatican.va/content/francesco/en/speeches/2019/september/documents/papa-francesco_20190927_eradigitale.html.

Francis. *Evangelii Gaudium* "On the Proclamation of the Gospel in Today's World" (November 24, 2013).

Francis. General audience. August 1, 2018. https://www.vatican.va/content/francesco/en/audiences/2018/documents/papa-francesco_20180801_udienza-generale.html.

Francis. General audience. November 29, 2023. https://www.vatican.va/content/

francesco/en/audiences/2023/documents/20231129-udienza-generale.html.

Francis. "Human. Meanings and Challenges," Address to the General Assembly of the Pontifical Academy for Life. February 12, 2024. https://www.vatican.va/content/francesco/en/speeches/2024/february/documents/20240212-pav.html.

Francis, *Laudate Deum* "On the Climate Crisis." October 4, 2023.

Francis. *Laudato si'* "On Care for Our Common Home." May 24, 2015.

Francis. *Message of the Holy Father for the 57th World Day of Peace on January 1, 2024.* December 14, 2023.

Francis. Message for the 58th World Day of Social Communications "Artificial Intelligence and the Wisdom of the Heart: Towards a Fully Human Communication" (January 24, 2024).

Freeman, Amy. "Remedies to Acedia in the Rhythm of Daily Life." In *Acedia.* Edited by Robert B. Kruschwitz. Waco, Texas: The Center for Christian Ethics at Baylor University, 2013. 36-44.

Fulay, Suyash, William Brannon, Shrestha Mohanty, Cassandra Overney, Elinor Poole-Dayan, Deb Roy, and Jad Kabbara. "On the Relationship between Truth and Political Bias in Language Models." arXiv (2024). DOI: 10.48550/ arxiv.2409.05283.

Gabriel, Iason, Arianna Manzini, Geoff Keeling, Lisa Anne Hendricks, Verena Rieser, Hasan Iqbal, Nenad Tomašev, et al. "The Ethics of Advanced AI Assistants" (2024): 103-4. https://doi.org/10.48550/arXiv.2404.16244.

Gallagher, Brian. "Does GPT-4 Really Understand What We're Saying." *Nautilus* (March 27, 2023). https://nautil.us/does-gpt-4-really-understand-what-were-saying-291034/.

Gerlich, Michael. "AI Tools in Society: Impacts on Cognitive Offloading and the Future of Critical Thinking." *Societies* 15, no. 6 (2025). https://doi.org/10.3390/ soc15010006.

Gerstein, Robert S. "Intimacy and Privacy." *Ethics* 89, no.1 (1978): 76. https://doi.org/10.1086/292105.

Gibbons, Kathleen Siobhan MacInnes. "Vice and Self Examination in the Christian Desert: An Intellectual Historical Reading of Evagrius Ponticus." Dissertation for the Doctor of Philosophy in Religion. Toronto: University of Toronto, 2011.

Glickman, Moshe and Tali Sharot, "How Human–AI Feedback Loops Alter Human Perceptual, Emotional and Social Judgements." *Nature Human Behavior* (2024). https://doi.org/10.1038/s41562-024-02077-2.

Glikson, Ella and Anita Williams Woolley. "Human Trust in Artificial Intelligence: Review of Empirical Research." *Academy of Management Annals* 14, no.2 (2020): 627–660. https://doi.org/10.5465/annals.2018.0057.

Goddard, Kate, Abdul Roudsari, and Jeremy C. Wyatt. "Automation Bias: A Systematic Review of Frequency, Effect Mediators, and Mitigators." *Journal of the American Medical Informatics Association* 19, vol.1 (2012): 121–127. https://doi.org/10.1136/amiajnl-2011-000089.

Goldstein, Josh A., Jason Chao, Shelby Grossman, Alex Stamos, and Michael Tomz. "How Persuasive is AI-generated Propaganda?" *PNAS Nexus* 3, no.2 (2024): 34. https://doi.org/10.1093/pnasnexus/pgae034.

Grant, George. "Thinking about Technology." In *Technology and Justice*. Notre Dame, Indiana: University of Notre Dame Press, 1986.

Granulo, A., A. Caprioli, C. Fuchs, and S. Puntoni. "Deployment of Algorithms in Management Tasks Reduces Prosocial Motivation." *Computers in Human Behavior*, 152 (2024): 108094. https://doi.org/10.1016/j.chb.2023.108094.

Green, Brian Patrick. "The Catholic Church and Technological Progress: Past, Present, and Future." *Religions* 8, no.106. https://doi.org/10.3390/rel8060106.

Grennan, Jillian and Roni Michaely. "Artificial Intelligence and High-Skilled Work: Evidence from Analysts." *Swiss Finance Institute Research Paper Series* No. 20-84. Geneva: Swiss Finance Institute, 2020. https://dx.doi.org/10.2139/ssrn.3681574.

Grossmann, Igor, Matthew Feinberg, Dawn C. Parker, Nicholas A. Christakis, Philip E. Tetlock, and William A. Cunningham. "AI and the Transformation of Social Science Research." *Science* 380 (2023): 1108–1109. https://doi.org/10.1126/science.adi1778.

Guardini, Romano. *Power And Responsibility: A Course Of Action For The New Age*. Washington, DC: Regnery, 1961.

Guardini, Romano. *The End of the Modern World*. Wilmington, DE: ISI Books, 1998.

Gulotta, Rebecca, William Odom, Jodi Forlizzi, and Haakon Faste. "Digital Artifacts as Legacy: Exploring the Lifespan and Value of Digital Data."

Proceedings of the SIGCHI Conference on Human Factors in Computing Systems (2013): 1813–1822. https://doi.org/10.1145/2470654.2466240.

Gunkel, David J. and Jordan Joseph Wales. "Debate: What Is Personhood in the Age of AI?" *AI and Society* 36 (2021). https://doi.org/10.1007/s00146-020-01129-1.

Haas, Lukas, Michal Skreta, Silas Alberti, and Chelsea Finn. "PIGEON: Predicting Image Geolocations" (2024). https://arxiv.org/abs/2307.05845.

Haenlein, Michael and Andreas Kaplan. "A brief history of artificial intelligence: On the past, present, and future of artificial intelligence." *California Management Review* 61, no. 4 (2019): 5-14. http://dx.doi.org/10.1177/0008125619864925.

Hagendorff, Thilo. "Deception Abilities Emerged in Large Language Models." *PNAS* 121, no.24 (2024): e2317967121. https://doi.org/10.1073/pnas.2317967121.

Hamilton, Eric and University of Florida. "What Makes Robots Feel Human? A New Scale Reveals the Secret." *Neuroscience News* (December 9, 2024). https://neurosciencenews.com/human-like-robots-neuroscience-28221/.

Hancock, P. A., Theresa T. Kessler, Alexandra D. Kaplan, John C. Brill, and James L. Szalma. "Evolving Trust in Robots: Specification Through Sequential and Comparative Meta-Analyses." *Human Factors: The Journal of the Human Factors and Ergonomics Society* 63, no.7 (2021): 1196–1229. https://doi.org/10.1177/0018720820922080.

Harrison, Joshua, Ehsan Toreini, and Maryam Mehrnezhad. "A Practical Deep Learning-Based Acoustic Side Channel Attack on Keyboards" (2023). https://arxiv.org/pdf/2308.01074.

Haslam, Nick. "Dehumanization: An Integrative Review." *Personal and Social Psychology Review* 10 (2006): 252–64. https://doi.org/10.1207/s15327957pspr1003_4.

He, Ai-Zhong and Yu Zhang. "AI-Powered Touch Points in the Customer Journey: A Systematic Literature Review and Research Agenda." *Journal of Research in Interactive Marketing* 17, no.4 (2023): 620-639. https://doi.org/10.1108/JRIM-03-2022-0082.

Hegel, Frank, Soren Krach, Tilo Kircher, Britta Wrede, and Gerhard Sagerer. "Understanding Social Robots: A User Study on Anthropomorphism." *Proceedings of the 17th IEEE International Symposium on Robot and Human*

Interactive Communication, (2008): 574-9. http://dx.doi.org/10.1109/RO-MAN.2008.4600728.

Hofmann, Valentin, Pratyusha Ria Kalluri, Dan Jurafsky, and Sharese King. "Dialect Prejudice Predicts AI Decisions about People's Character, Employability, and Criminality" (2024). https://doi.org/10.48550/arXiv.2403.00742.

Holt, Jack, James Nicholson, and Jan David Smeddinck. "From Personal Data to Digital Legacy: Exploring Conflicts in the Sharing, Security and Privacy of Post-mortem Data." *Proceedings of the Web Conference 2021 (WWW '21)*: 2745–56. https://doi.org/10.1145/3442381.3450030.

Hoppe, Sabrina, Tobias Loetscher, Stephanie A. Morey, and Andreas Bulling. "Eye Movements During Everyday Behavior Predict Personality Traits." *Frontiers in Human Neuroscience* 12 (2018). https://doi.org/10.3389/fnhum.2018.00105.

Horkheimer, Max and Theodor W. Adorno. *Dialectic of Enlightenment.* Stanford, California: Stanford University Press, 2002.

Horning, Rob. "After the Sunsets." Internal Exile blog (October 15, 2024). https://robhorning.substack.com/p/after-the-sunsets.

Horning, Rob. "Artificial Intentionality." Internal Exile blog (September 20, 2024). https://robhorning.substack.com/p/artificial-intentionality.

Hu, T., Y. Kyrychenko, S. Rathje, et al. "Generative Language Models Exhibit Social Identity Biases." *Nature Computer Science* 5 (2025): 65–75. https://doi.org/ 10.1038/s43588-024-00741-1.

Hyde, Steven J., Eric Bachura, Jonathan Bundy, Richard T. Gretz, and William Gerard Sanders. "The Tangled Webs We Weave: Examining the Effects of CEO Deception on Analyst Recommendations." *Strategic Management Journal* 45, no.1 (2024): 66-112. http://doi.org/10.1002/smj.3546.

IBM Institute for Business Value. *Augmented Work for an Automated, AI-Driven World.* Armonk, New York: IBM, 2023. https://www.ibm.com/downloads/cas/NGAWMXAK.

Jackson, J. C., K. C. Yam, P. M. Tang, C. G. Sibley, and A. Waytz. "Exposure to Automation Explains Religious Declines." *Proceedings of the National Academy of Sciences* 120: e2304748120 (2023).

Jakesch, Maurice, Jeffrey T. Hancock, and Mor Naaman. "Human Heuristics for AI-Generated Language Are Flawed." *Proceedings of the National Academy of Sciences* 120, no.11 (2023): e2208839120.

https://doi.org/10.1073/pnas.2208839120.

Ji, Ziwei, Nayeon Lee, Rita Frieske, Tiezheng Yu, et al. "Survey of Hallucination in Natural Language Generation." *ACM Computing Surveys* 55, no.12 (Mar. 2023): 1-38. https://doi.org/10.1145/3571730.

Jiang, Ray, Silvia Chiappa, Tor Lattimore, András György, and Pushmeet Kohli. "Degenerate Feedback Loops in Recommender Systems." *Proceedings of the 2019 AAAI/ACM Conference on AI, Ethics, and Society* (2019): 383–390. https://dl.acm.org/doi/10.1145/3306618.3314288.

John Paul II. "A Fundamental Disquiet in All Human Existence," General Audience (May 28, 1980).

John Paul II. Address to the Thirty-Fifth General Assembly of the World Medical Association (October 29, 1983).

John Paul II. *Centesimus annus* "On the Hundredth Anniversary of *Rerum Novarum.*" May 1, 1991.

John Paul II. "Creation as a Fundamental and Original Gift," General Audience (January 2, 1980).

John Paul II. *Crossing the Threshold of Hope.* Edited by Vittorio Messori. New York: Alfred A. Knopf, 2005.

John Paul II. "Dominion over the Other in the Interpersonal Relation," General Audience (June 18, 1980).

John Paul II, *Evangelium vitae* "The Gospel of Life." March 25, 1995.

John Paul II. *Familiaris Consortio* "On the Role of the Christian Family in the Modern World." November 22, 1981.

John Paul II. *Fides et ratio* "On the Relationship between Faith and Reason." September 14, 1998.

John Paul II. *Gratissimam sane* "Letter to Families." Feb. 2, 1994.

John Paul II. Homily, Closing of World Youth Day (August 20, 2000). https://www.vatican.va/content/john-paul-ii/en/homilies/2000/documents/hf_jp-ii_hom_20000820_gmg.html.

John Paul II. "Real Significance of Original Nakedness," General Audience (December 12, 1979).

John Paul II. *Redemptor hominis.* March 4, 1979.

John Paul II. "Relationship of Lust to Communion of Persons," General Audience (June 4, 1980).

John Paul II. *The Theology of the Body: Human Love in the Divine Plan.* Boston:

Pauline Books and Media, 1997.

John Paul II. *Veritatis splendor* "The Splendor of Truth." August 6, 1993.

Joinson, A. N., U.-D. Reips, T. Buchanan, and C. B. P. Schofield. "Privacy, Trust, and Self-Disclosure Online." *Human-Computer Interaction*, 25, no.1 (2010): 1–24.

Jones, Christopher D. "The Problem of Acedia in Eastern Orthodox Morality." *Studies in Christian Ethics* 33, no.3 (2020): 336-51.

Jones, Jeffrey M. "Church Attendance Has Declined in Most U.S. Religious Groups." Gallup (March 25, 2024). https://news.gal-lup.com/poll/642548/church-attendance-declined-religious-groups.aspx.

Jurgens, Jeremy. Press conference (January 18, 2023). World Economic Forum. https://www.weforum.org/meetings/world-economic-forum-annual-meet-ing-2023/sessions/press-conference-global-cybersecurity-outlook-2023/.

Kalberg, Stephen. "Max Weber's Types of Rationality: Cornerstones for the Analysis of Rationalization Processes in History." *The American Journal of Sociology*, 85, no. 5 (1980): 1145-1179.

Kanazawa, Kyogo, Daiji Kawaguchi, Hitochi Shigeoka, and Yasutora Watanabe. "AI Skill and Productivity: The Case of Taxi Drivers." *IZA Discussion Papers*. Bonn: Institute of Labor Economics IZA, 2022: 15677. https://hdl.handle.net/10419/267414.

Kapoor, Sayash and Arvish Narayanan. "Leakage and the Reproducibility Crisis in Machine-Learning-Based Science." *Patterns* 4, no.9 (2023): 100804. https://doi.org/10.1016/j.patter.2023.100804.

Karter, Erin. "As Newspapers Close, Struggling Communities Are Hit Hardest by the Decline in Local Journalism." *Northwestern Now* (June 29, 2022). https://news.northwestern.edu/stories/2022/06/newspapers-close-decline-in-local-journalism/.

Kenton, Zachary, Tom Everitt, Laura Weidinger, Iason Gabriel, Vladimir Mikulik, and Geoffrey Irving. "Alignment of Language Agents" (2021). http://arxiv.org/abs/2103.14659.

Kidd, Celeste and Abeba Birhane. "How Generative AI Models Can Distort Human Beliefs." *Science* 380, no.6651 (2023): 1222-1223. https://doi.org/10.1126/science.adi0248.

Kim, Hye-young and Ann L. McGill. "AI-Induced Dehumanization." *Journal of Communication Psychology* (2024). DOI: 10.1002/jcpy.1441.

Kim, TaeWoo, Hyejin Lee, Michelle Yoosun Kim, SunAh Kim, and Adam Duhachek. "AI Increases Unethical Consumer Behavior Due To Reduced Anticipatory Guilt." *Journal of the Academy of Marketing Science* 51 (2022): 785–801. http://dx.doi.org/10.1007/s11747-021-00832-9.

King, Ryan Erik. "Waymo Is Suing 2 Alleged Vandals For Over $270,000 In Damages To Its Robotaxis." *Jalopnik* (July 24, 2024). https://jalopnik.com/waymo-is-suing-2-alleged-vandals-for-over-270-000-in-d-1851603716.

Kittler, Friedrich A. *Gramophone, Film, Typewriter.* Translated by Geoffrey Winthrop-Young and Michael Wutz. Stanford: Stanford University Press, 1999. 200-208.

Klym, Natalie. "The Technologists are Not in Control: What the Internet Experience Can Teach us about AI Ethics and Responsibility." In *The State of AI Ethics Report,* Volume 6. Montreal AI Ethics Institute (January 2022). https://montrealethics.ai/wp-content/uploads/2022/01/State-of-AI-Ethics-Report-Volume-6-February-2022.pdf.

Kolodny, Niko and John Brunero. "Instrumental Rationality." *The Stanford Encyclopedia of Philosophy*, Summer 2023 Edition. Edited by Edward N. Zalta & Uri Nodelman. https://plato.stanford.edu/archives/sum2023/entries/rationality-instrumental.

Korsgaard, Christine. "The Normativity of Instrumental Reason." In *Ethics and Practical Reason.* Edited by Garrett Cullity and Berys Gaut. Oxford: Clarendon Press, 1997.

Kroger, Jacob Leon, Otto Hans-Martin Lutz, and Florian Muller. "What Does Your Gaze Reveal About You? On the Privacy Implications of Eye Tracking." In *Privacy and Identity Management.* Edited by Michael Friedewald, Melek Onen, Eva Lievens, Stephan Krenn, and Samuel Fricker. Berlin: Springer, 2020. https://doi.org/10.1007/978-3-030-42504-3_15.

Kyriacou, Christos. "Artificial Moral Intelligence and Computability: An Aristotelian Perspective." *AI and Ethics* (2024). https://doi.org/10.1007/s43681-024-00543-1.

Laestadius, Linnea, Andrea Bishop, Michael Gonzalez, Diana Illenčík, and Celeste Campos-Castillo. "Too Human and Not Human Enough: A Grounded Theory Analysis of Mental Health Harms from Emotional Dependence on the Social Chatbot Replika." *New Media and Society* (2022).

https://doi.org/10.1177/14614448221142007.

Lai, Peihua and Stephen Westland. "Machine Learning for Colour Palette Extraction from Fashion Runway Images." *International Journal of Fashion Design, Technology and Education* 13, no.3 (2020): 334-340. https://doi.org/10.1080/17543266.2020.1799080.

Lannoy, Valérie. "AI Based Plagiarism Detectors: Plagiarism Fighters or Makers." *Medical Writing* 32, no.3 (2023): 44-7. doi: 10.56012/ovnr4109.

Lee, Dokyun and Kartik Hosanagar. "How Do Recommender Systems Affect Sales Diversity? A Cross-Category Investigation via Randomized Field Experiment." *Information Systems Research* 30, no.1 (2019): 239–59.

Lee, Jin Pyo, Hanhyeok Jang, Yeonwoo Jang, Hyeonseo Song, Suwoo Lee, Pooi See Lee, and Jiyun Kim. "Encoding of Multi-Modal Emotional Information via Personalized Skin-Integrated Wireless Facial Interface." *Nature Communications* 15 (2024): 530. https://doi.org/10.1038/s41467-023-44673-2.

Legg, Shane and Marcus Hutter. "A Collection of Definitions of Intelligence." *Frontiers in Artificial intelligence and Applications* 157 (2007): 17-24. https://doi.org/10.48550/arXiv.0706.3639.

Lehman, Joel, Jeff Clune, Dusan Misevic, Christoph Adami, Lee Altenberg, Julie Beaulieu, Peter J. Bentley, et al. "The Surprising Creativity of Digital Evolution: A Collection of Anecdotes from the Evolutionary Computation and Artificial Life Research Communities." *Artificial Life* 26, no.2 (2020): 274–306. https://doi.org/10.1162/artl_a_00319.

Lehmann, Sune. "Using Sequences of Life-Events to Predict Human Lives." *Nature Computational Science* 4 (2023): 43-56. https://doi.org/10.1038/s43588-023-00573-5.

Leuenberger, Muriel. "AI 'Can Stunt the Skills Necessary for Independent Self-Creation': Relying on Algorithms Could Reshape Your Entire Identity Without You Realizing." *Live Science* (October 27, 2024). https://www.livescience.com/ technology/artificial-intelligence/ai-can-stunt-the-skills-necessary-for-independent-self-creation-relying-on-algorithms-could-reshape-your-entire-identity-without-you-realizing.

Leung, King-Ho. "The Picture of Artificial Intelligence and the Secularization of Thought." *Political Theology* 20, no.6 (2019): 463. https://doi.org/10.1080/1462317X.2019.1605725.

Li, Peiyao, Noah Castelo, Zsolt Katona, and Miklos Sarvary. "Frontiers:

Determining the Validity of Large Language Models for Automated Perceptual Analysis." *Marketing Science* 43, no.2 (2024): 254-6. https://doi.org/10.1287/mksc.2023.0454.

Liang, Weixin, Zachary Izzo, Yaohui Zhang, Haley Lepp, Hancheng Cao, Xuandong Zhao, Lingjiao Chen, et al. "Monitoring AI-Modified Content at Scale: A Case Study on the Impact of ChatGPT on AI Conference Peer Reviews" (2024). https://arxiv.org/pdf/2403.07183.

Liang, Weixin, Mert Yuksekgonul, Yining Mao, Eric Wu, and James Zou. "GPT Detectors Are Biased against Non-Native English Writers." *Patterns* 4, no.7 (2023): 100779. https://doi.org/10.1016/j.patter.2023.100779.

Lu, Ywien and Cade Metz. "Cruise's Driverless Taxi Service in San Francisco Is Suspended." *New York Times* (October 24, 2023). https://www.nytimes.com/2023/10/24/technology/cruise-driverless-san-francisco-suspended.html.

Macy, Michael W. and Robert Willer. "From Factors to Actors: Computational Sociology and Agent-Based Modeling." *Annual Review of Sociology* 28 (2002): 143–166. doi: 10.1146/annurev.soc.28.110601.141117.

Manyika, James, Jake Silberg, and Brittany Presten. "What Do We Do about the Biases in AI?" *Harvard Business Review* (October 25, 2019). https://hbr.org/2019/10/what-do-we-do-about-the-biases-in-ai.

Marche, Stephen. "Welcome to the Big Blur." *The Atlantic* (March 14, 2023). https://www.theatlantic.com/technology/archive/2023/03/gpt4-arrival-human-artificial-intelligence-blur/673399e.

Marriott, Hannah R. and Valentina Pitardi. "One Is the Loneliest Number … Two Can Be as Bad as One. The Influence of AI Friendship Apps on Users' Well-Being and Addiction." *Psychology and Marketing* 41, no.1 (2024): 86-101. https://doi.org/10.1002/mar.21899.

Martin, Randy. "Anomie, Spirituality, and Crime." *Journal of Contemporary Criminal Justice*, 16, no.1 (2000): 75-98. https://doi.org/10.1177/1043986200016001005.

Matz, S. C., J. D. Teeny, S. S. Vaid, H. Peters, G. M. Harari, and M. Cerf. "The Potential of Generative AI for Personalized Persuasion at Scale." *Scientific Reports* 14, no. 4692 (2024). https://doi.org/10.1038/s41598-024-53755-0.

McCarthy, John, M. L. Minsky, Nathaniel Rochester, C. E. Shannon. "A Proposal for the Dartmouth Summer Research Project on Artificial

Intelligence" (1955). http://www-formal.stanford.edu/jmc/history/dart-mouth/dartmouth.html.

McDonald, Matthew. "Catholic Answers Pulls Plug on 'Father Justin' AI Priest." *National Catholic Register* (April 24, 2024). https://www.ncregister.com/news/catholic-answers-ai-priest-cancelled.

McLuhan, Marshall. *Understanding Media: The Extensions of Man.* McGraw-Hill, 1964.

Megaw, Nicholas. "Investors use AI to glean signals behind executives' soothing words." *Financial Times* (Nov. 12, 2023). https://www.ft.com/content/ee2788dd-aca5-4214-8a08-d88081eac1b9.

Meinke, Alexander, Bronson Schoen, Jérémy Scheurer, Mikita Balesni, Rusheb Shah, and Marius Hobbhahn. "Frontier Models are Capable of In-context Scheming." arXiv (January 14, 2025). https://arxiv.org/abs/2412.04984.

Mental Health America. *State of Mental Health in America* (2024). https://mhanational.org/issues/state-mental-health-america.

Merrill, Kelly Jr., Jihyun Kim, and Chad Collins. "AI Companions for Lonely Individuals and the Role of Social Presence." *Communication Research Reports* 39, no.2 (2022): 93–103. https://doi.org/10.1080/08824096.2022.2045929.

Microsoft. "Copilot+ PC Features." https://www.microsoft.com/en-us/windows/copilot-plus-pcs?r=1#faq2.

Miller, Elizabeth J., Ben A. Steward, Zak Witkower, Clare A. M. Sutherland, Eva G. Krumhuber, and Amy Dawel. "AI Hyperrealism: Why AI Faces Are Perceived as More Real Than Human Ones." *Psychological Science* 34, no.12 (2023): 1390-1403. https://doi.org/10.1177/09567976231207095.

Melanie Mitchell. "The Metaphors of Artificial Intelligence." *Science* 386, no. 6723 (2024). https://www.science.org/doi/10.1126/science.adt6140.

Mlonyeni, Philip Maxwell Thingbø. "Personal AI, Deception, and the Problem of Emotional Bubbles." *AI and Society* 10 (2024). https://doi.org/10.1007/s00146-024-01958-4.

Morewedge, C. K., S. Mullainathan, H. F. Naushan, C. R. Sunstein, J. Kleinberg, M. Raghavan, and J. O. Ludwig. "Human Bias in Algorithm Design." *Nature Human Behaviour* 7, no.11 (2023).

Mori, Masahiro, Karl F. MacDorman, and Norri Kageki. "The Uncanny Valley [From the Field]." *IEEE Robotics & Automation Magazine*, 19, no.2 (2012):

98–100. https://doi.org/10.1109/MRA.2012.2192811.

Morris, Meredith Ringel and Jed R. Brubaker. "Generative Ghosts: Anticipating Benefits and Risks of AI Afterlives" (2024). https://arxiv.org/html/2402.01662v1.

Mou, Yi and Kun Xu. "The Media Inequality: Comparing the Initial Human-Human and Human-AI Social Interactions." *Computers in Human Behavior* 72 (2017): 432–40. https://doi.org/10.1016/j.chb.2017.02.067.

Mouhsine, Wael. "Machine Learning Helps Uncover Hidden Consumer Motivations." *Tech Xplore* (December 2, 2024). https://techxplore.com/news/2024-12-machine-uncover-hidden-consumer.html.

Mozilla Foundation. "Romantic AI." February 7, 2024. https://foundation.mozilla.org/en/privacynotincluded/romantic-ai/.

Nabavi, Ehsan. "AI Could Crack Unsolvable Problems — and Humans Won't Be Able to Understand the Results." *Live Science* (January 5, 2025). https://www.livescience.com/technology/artificial-intelligence/ai-could-crack-unsolvable-problems-and-humans-wont-be-able-to-understand-the-results.

National Association of Insurance Commissioners. *Private Passenger Auto Artificial Intelligence/Machine Learning Survey Results.* Washington, DC: NAIC, 2022. https://content.naic.org/sites/default/files/inline-files/PP%20Auto%20Survey%20Team%20Report%20120822.pdf.

National Association of Insurance Commissioners. *Life Insurance Artificial Intelligence/Machine Learning Survey Results.* Washington, DC: NAIC, 2023. https://content.naic.org/sites/default/files/inline-files/life-ai-survey-report-final.pdf.

Nault, Jean-Charles. "Acedia: Enemy of Spiritual Joy." *Communio* 31 (2004): 236-259.

Nault, Jean-Charles. *The Noonday Devil: Acedia, the Unnamed Evil of Our Times.* Translated by Michael J. Miller. San Francisco: Ignatius Press, 2013.

Nightingale, Sophie J. and Hany Farid. "AI-Synthesized Faces are Indistinguishable from Real Faces and More Trustworthy." *PNAS* 119, no.8 (2022): e2120481119. https://doi.org/10.1073/pnas.2120481119.

Nishant, Rohit, Dirk Schneckenberg, and M.N. Ravishankar. "The Formal Rationality of Artificial Intelligence-Based Algorithms and the Problem of Bias." *Journal of Information Technology* 39, no.1 (2024): 19-40.

https://doi.org/10.1177/02683962231176842.

Noy, Shakked and Whitney Zhang. "Experimental Evidence on the Productivity Effects of Generative Artificial Intelligence." *Science* 381, no.6654 (2023): 187-192. https://doi.org/10.1126/science.adh2586.

Ockholm, Dennis. "Staying Put to Get Somewhere." In *Acedia*. Ed. Robert B. Kruschwitz. Waco, Texas: The Center for Christian Ethics at Baylor University, 2013. 19-25.

Onur, A., A. Seidmann, B. Gu, and N. Mazar. "The Effect of Interpretable AI on Repetitive Managerial Decision-Making under Uncertainty." Research paper. Boston: Boston University Questrom School of Business, 2023. https://doi.org/10.2139/ssrn.4331145.

OpenAI. *OpenAI o1 System Card* (December 5, 2024). https://cdn.openai.com/o1-system-card-20241205.pdf.

Parisi, Lorenza and Francesca Comunello. "Dating in the Time of 'Relational Filter Bubbles': Exploring Imaginaries, Perceptions and Tactics of Italian Dating App Users." *The Communication Review* 23, no. 1 (2020): 66–89. https://doi.org/10.1080/10714421.2019.1704111.

Joon Sung Park, et al. "Generative Agent Simulations of 1,000 People." arXiv (2024). DOI: 10.48550/arxiv.2411. 10109.

Park, Joon Sung, Joseph C. O'Brien, Carrie J. Cai, Meredith Ringel Morris, Percy Liang, and Michael S. Bernstein. "Generative Agents: Interactive Simulacra of Human Behavior" (2023). https://doi.org/10.48550/arXiv.2304.03442.

Park, Peter S., Simon Goldstein, Aidan O'Gara, Michael Chen, and Dan Hendrycks. "AI Deception: A Survey of Examples, Risks, and Potential Solutions." *Patterns* 5, no.5 (2024): 100988. https://doi.org/10.1016/j.patter.2024.100988.

Paul, Andrew. "A Crowd Torched a Waymo Robotaxi in San Francisco." Popular Science (February 12, 2024). https://www.popsci.com/technology/waymo-torched-vandals.

Paul VI. *Humanae vitae* "On the Regulation of Birth." July 25, 1968.

Pellert, Max, Clemens Lechner, Claudia Wagner, Beatrice Rammstedt, and Markus Strohmaier. "AI Psychometrics: Assessing the Psychological Profiles of Large Language Models through Psychometric Inventories." *Perspectives on Psychological Science* (2024). https://doi.org/10.1177/17456916231214460.

Perez, Carlos E. "12 Blind Spots in AI Research." *Medium* (December 25, 2018). https://medium.com/intuitionmachine/12-blind-spots-about-human-cognition-1883d0d58e0a.

Peters, Heinrich and Sandra C. Matz. "Large Language Models Can Infer Psychological Dispositions of Social Media Users." *PNAS Nexus* 3 (2024). https://doi.org/10.1093/pnasnexus/pgae231

Peters, Uwe and Mary Carman. "Cultural Bias in Explainable AI Research: A Systematic Analysis." *Journal of Artificial Intelligence Research* 79 (2024): 971-1000. https://doi.org/10.1613/jair.1.14888.

Pettersen, Lene & Runar Døving. "The Construction of Matches on Dating Platforms." *Nordic Journal of Science and Technology Studies* 11, no.1 (2023): 13-27.

Pew Research. "Religious 'Nones' in America: Who They Are and What They Believe" (January 24, 2024). https://www.pewresearch.org/religion/2024/01/24/religious-nones-in-america-who-they-are-and-what-they-believe/.

Phillips, Pete. "Why Your Bible App's 'Verse of the Day' Feature Could Be Skewing Your View of God." *Premier Christianity* (October 2, 2018). https://www.premierchristianity.com/home/why-your-bible-apps-verse-of-the-day-feature-could-be-skewing-your-view-of-god/3511.article.

Pieper, Josef. *Leisure: The Basis of Culture*. South Bend, Indiana: St. Augustine's Press, 1998.

Pitardi, V. and H. Marriott. "Alexa, She's Not Human But . . . Unveiling the Drivers of Consumers' Trust in Voice-Based Artificial Intelligence." *Psychology and Marketing*, 38, no.4 (2021): 626–42.

Postman, Neil. *Technopoly: The Surrender of Culture to Technology*. New York: Random House, 1993.

Poursabzi-Sangdeh, Forough, Daniel G Goldstein, Jake M Hofman, Jennifer Wortman Wortman Vaughan, and Hanna Wallach. "Manipulating and Measuring Model Interpretability." *CHI '21: Proceedings of the 2021 CHI Conference on Human Factors in Computing Systems* (2021): 237. https://dl.acm.org/doi/10.1145/3411764.3445315.

Quay, Grayson. "Algorithmic Spirituality." *First Things* blog (June 7, 2024). https://www.firstthings.com/web-exclusives/2024/06/algorithmicspirituality.

Quinn, Warren. "Putting Rationality in its Place." In *Morality and Action*. Cambridge, Massachusetts: Cambridge University Press, 1993. 228-55.

Quiroz-Gutierrez, Marco. "Elon Musk Says There's a 10% to 20% Chance that AI 'Goes Bad,' Even While He Raises Billions for His Own Startup xAI." *Fortune* (October 30, 2024). https://fortune.com/2024/10/30/elon-musk-ai-could-go-bad-existential-threat-xai-fundraising/.

Radivojevic, Kristina, Nicholas Clark, and Paul Brenner. "LLMs Among Us: Generative AI Participating in Digital Discourse" (2024). https://doi.org/10.48550/arXiv.2402.07940.

Ratzinger, Joseph. *A Turning Point for Europe? The Church in the Modern World: Assessment and Forecast*, 2nd ed. Translated by Brian McNeil. San Francisco: Ignatius Press, 1994.

Ratzinger, Joseph. *The God of Jesus Christ: Meditations on the Triune God*. Translated by Brian McNeil. San Francisco: Ignatius Press, 2008.

Ratzinger, Joseph. "Truth and Freedom." *Communio* 23 (1996): 16–35.

Ray, Julie and Gallup. "Americans Express Real Concerns About Artificial Intelligence" (August 27, 2024). https://news.gallup.com/poll/648953/americans-express-real-concerns-artificial-intelligence.aspx.

Reilly, Christopher M. "How Artificial Intelligence Technology Encourages the Vice of Acedia." *Divus Thomas* 126, no. 2 (2023): 152-175.

Reilly, Christopher M. "Technological Domination: Its Moral Significance in Bioethics." *National Catholic Bioethics Quarterly* 23, no. 1 (Spring 2023): 23–35.

RenAIssance Foundation. "AI Ethics for Peace – Hiroshima, July 9, 2024." https://www.romecall.org/ai-ethics-for-peace-hiroshima-july-9th-2024/.

Rheu, Minjin, Ji Youn Shin, Wei Peng, and Jina Huh-Yoo. "Systematic Review: Trust-Building Factors and Implications for Conversational Agent Design." *International Journal of Human–Computer Interaction* 37, no.1 (2021): 81–96. https://doi.org/10.1080/10447318.2020.1807710.

Rio-Chanona, Maria del, Nadzeya Laurentsyeva, and Johannes Wachs. "Are Large Language Models a Threat to Digital Public Goods? Evidence from Activity on Stack Overflow" (2023). https://arxiv.org/abs/2307.07367.

Roberts, John, Max Baker, and Jane Andrew. "Artificial Intelligence and Qualitative Research: The Promise and Perils of Large Language Model (LLM) 'Assistance'." *Critical Perspectives on Accounting* 99 (2024): 102722.

https://doi.org/10.1016/j.cpa.2024.102722.

Rosenblatt, Matthew, Link Tejavibulya, Rongtao Jiang, Stephanie Noble, and Dustin Scheinost. "Data Leakage Inflates Prediction Performance in Connectome-Based Machine Learning Models." *Nature Communications* 15 (2024): 1829. https://doi.org/10.1038/s41467-024-46150-w.

Rossignac-Milon, Maya, Niall Bolger, Katherine S. Zee, Erica J. Boothby, and E. Tory Higgins. "Merged Minds: Generalized Shared Reality in Dyadic Relationships." *Journal of Personality and Social Psychology* 120, no.4 (2021): 882–911. http://dx.doi.org/10.1037/pspi0000266.

Rudinger, Rachel, Jason Naradowsky, Brian Leonard, and Benjamin Van Durme. "Gender Bias in Coreference Resolution." *Proceedings of NAACL-HLT 2018* (2018): 8–14. https://doi.org/10.18653/v1/N18-2002.

Russell, Stuart and Peter Norvig. *Artificial Intelligence: A Modern Approach*, 4th edition. Hoboken: Pearson, 2021.

Sacasas, L. M. "Re-Sourcing the Mind." *The Convivial Society* 5, no.9 (August 1, 2024). https://theconvivialsociety.substack.com/p/re-sourcing-the-mind.

Salvi, Francesco, Manoel Horta Ribeiro, Riccardo Gallotti, and Robert West. "On the Conversational Persuasiveness of Large Language Models: A Randomized Controlled Trial" (2024). https://arxiv.org/abs/2403.14380.

Savcisens, Germans, Tina Eliassi-Rad, Lars Kai Hansen, Laust Hvas Mortensen, Lau Lilleholt, Anna Rogers, Ingo Zettler, and Sune Lehmann. "Using Sequences of Life-Events to Predict Human Lives." *Nature Computational Science* 4 (2024): 43-56. https://doi.org/10.1038/s43588-023-00573-5.

Scherz, Paul. "Data Ethics, AI, and Accompaniment: The Dangers of Depersonalization in Catholic Health Care." *Theological Studies* 83, no.2 (2022): 271-292. DOI: 10.1177/00405639221096770.

Schindler, David L. "America's Technological Ontology and the Gift of the Given: Benedict XVI on the Cultural Significance of the Quaerere Deum." *Communio* 38.2 (2011): 237-278.

Sen, Amartya. *Rationality and Freedom*. Cambridge, Massachusetts: Harvard University Press, 2002.

Seymour, Kiley, Jarrod McNicoll, and Roger Koenig-Robert. "Big Brother: The Effects of Surveillance on Fundamental Aspects of Social Vision." *Neuroscience of Consciousness*, 1 (2024): niae039. https://doi.org/10.1093/nc/niae039.

Shadle, Matthew. "Exposed Before Digital Omniscience: A Theological Reading of Surveillance Capitalism." *Church Life Journal* (August 3, 2021). https://churchlifejournal.nd.edu/articles/surveillance-capitalism-in-a-post-privacy-age.

Shanahan, Murray. "Talking about Large Language Models." arXiv (February 16, 2023). https://doi.org/10.48550/arXiv.2212.03551.

Sharkey, A. and N. Sharkey. "Granny and the Robots: Ethical Issues in Robot Care for the Elderly." *Ethics of Information Technology* 14 (2022): 27–40. https://doi.org/10. 1007/s10676- 010- 9234-6.

Sharkey, A. and N. Sharkey. "We Need to Talk about Deception in Social Robotics!" *Ethics of Information Technology* 23, no.3 (2021): 309–316. https://doi.org/10.1007/s10676- 020- 09573-9.

Shavit, Yonadav, Sandhini Agarwal, Miles Brundage, Steven Adler, Cullen O'Keefe, Rosie Campbell, Teddy Lee, et al. "Practices for Governing Agentic AI Systems" (2023). https://cdn.openai.com/papers/practices-for-governing-agentic-ai-systems.pdf.

Sidoti, Olivia, Eugenie Park, and Jeffrey Gottfried. "About a Quarter of U.S. Teens Have Used ChatGPT for Schoolwork – Double the Share in 2023." Pew Research Center (January 15, 2025). https://www.pewresearch.org/short-reads/2025/01/15/about-a-quarter-of-us-teens-have-used-chatgpt-for-schoolwork-double-the-share-in-2023/.

Skjuve, Marita, Asbjørn Følstad, Knut Inge Fostervold, and Petter Bae Brandtzaeg. "My Chatbot Companion - a Study of Human-Chatbot Relationships." *International Journal of Human-Computer Studies* 149 (2021): 102601. https://doi.org/10.1016/j.ijhcs.2021.102601.

Skjuve, Marita, Asbjorn Følstad, Knut Inge Fostervold, and Petter Bae Brandtzaeg. "A Longitudinal Study of Human–Chatbot Relationships." *International Journal of Human-Computer Studies* 168 (2022): 102903. https://doi.org/10.1016/j.ijhcs.2022.102903.

Smith, Brian Cantwell. *The Promise of Artificial Intelligence: Reckoning and Judgment.* Cambridge, Massachusetts: MIT Press, 2019.

Smith, S. E. "What Happens When the Internet Disappears?" *The Verge* (December 18, 2024). https://www.theverge.com/24321569/internet-decay-link-rot-web-archive-deleted-culture.

Soriano, Ashley. "Recycling Trucks with AI-Powered Cameras Raise Privacy

Concerns." *NewsNation* (October 20, 2024). https://www.newsnation-now.com/business/tech/ai/recycling-artificial-intelligence-cameras-privacy.

Spielthenner, Georg. "Instrumental Reasoning Reconsidered." *European Journal of Analytic Philosophy* 4, no. 1 (2008): 59-76.

Spitale, Giovanni, Nikola Biller-Andorno, and Federico Germani. "AI Model GPT-3 (Dis)Informs Us Better than Humans." *Science Advances* 9, no.26 (2023). https://doi.org/10.1126/sciadv.adh1850.

Staab, Robin, Mark Vero, Mislav Balunović, and Martin Vechev. "Beyond Memorization: Violating Privacy Via Inference with Large Language Models" (2024). https://doi.org/10.48550/arXiv.2310.07298.

STM Association. *AI Ethics in Scholarly Communication: STM Best Practice Principles for Ethical, Trustworthy and Human-Centric AI.* Oxford: STM, 2021. https://www.stm-assoc.org/2021_04_29_STM_AI_White_Paper_April2021.pdf.

Stokel-Walker, Chris. "Over 70 Per Cent of Students in US Survey Use AI for School Work." *New Scientist* (December 13, 2024). https://www.newscientist.com/article/2460254-over-70-per-cent-of-students-in-us-survey-use ai-for-school-work/

Sultana, Mark. "Combatting Acedia: The Neptic Attitude." *Heythrop Journal*, 63, no. 4 (2019): 828-844.

Sweeney, M. E., and E. Davis. "Alexa, Are You Listening?" *Information Technology and Libraries*, 39, no.4 (2021). https://doi.org/10.6017/ital.v39i4.12363.

Taylor, Charles. *The Sources of the Self: The Making of the Modern Identity.* Cambridge, Massachusetts: Harvard University Press, 1989.

Taylor, Michael Dominic. "'Riveted with Faith unto Your Flesh': Technology's Flight from Actuality and the Word Made Flesh." *Communio* 49 (2022).

Terán-Somohano, Alejandro. "AI Deadbots and the Need for Christian Hope." Word on Fire (June 14, 2024). https://www.wordonfire.org/articles/ai-deadbots-and-the-need-for-christian-hope.

Terán-Somohano, Alejandro. "The Banalization of Intelligence." Word on Fire (August 13, 2024). https://www.wordonfire.org/articles/the-banalization-of-intelligence/.

To, Christopher, Dylan Wiwad, and Maryam Kouchaki. "Economic Inequality Reduces Sense of Control and Increases the Acceptability of Self-Interested

Unethical Behavior." *Journal of Experimental Psychology* 152, no.10 (2023): 2747-74. https://psycnet.apa.org/doi/10.1037/xge0001423.

Unger, Moshe, Michel Wedel, and Alexander Tuzhilin. "Predicting Consumer Choice from Raw Eye-Movement Data Using the RETINA Deep Learning Architecture." *Data Mining and Knowledge Discovery* 38 (2024): 1069-1100. https://doi.org/10.1007/s10618-023-00989-7.

University of Cambridge. "AI's Next Frontier: Selling Your Intentions before You Know Them." *Tech Xplore* (December 29, 2024). https://techxplore.com/news/2024-12-ai-frontier-intentions.amp.

Valenzuela, Ana, et al., "How Artificial Intelligence Constrains the Human Experience." *Journal of the Association for Consumer Research* 9, no.3 (2024). https://doi.org/10.1086/730709.

Vallor, Shannon. "The Danger Of Superhuman AI Is Not What You Think." *Noema Magazine* (May 23, 2024). https://www.noemamag.com/the-danger-of-superhuman-ai-is-not-what-you-think/.

Vega, Pedro. "Are AI-Generated Homilies Suitable for the Edification and Flourishing of the Catholic Faithful?" *Homiletic and Pastoral Review.* May 24, 2024. https://www.hprweb.com/2024/05/are-ai-generated-homilies-suitable-for-the-edification-and-flourishing-of-the-catholic-faithful/.

Veluri, Bandhav, Malek Itani, Tuochao Chen, Takuya Yoshioka, and Shyamnath Gollakota. "Look Once to Hear: Target Speech Hearing with Noisy Examples." *CHI '24: Proceedings of the CHI Conference on Human Factors in Computing Systems* (2024), 37: 1-16. https://doi.org/10.1145/3613904.3642057.

Ventre-Dominey, J., G. Gibert, M. Bosse-Platiere, A. Farnè, P. F. Dominey, and F. Pavani. "Embodiment into a Robot Increases Its Acceptability." *Scientific Reports* 9, 10083 (2019). https://doi.org/10.1038/s41598-019-46528-7.

Vilvang, Chrys. "Between Automated Memory and History: Blocking 'Sensitive Locations' from Apple Memories." *Memory, Mind & Media* 3 (2024): e8. https://doi.org/10.1017/mem.2024.4.

Vocca, Riccardo. "Generative AI Makes You More Creative, But It Makes Us All Less So." Intelligent Friend blog (December 8, 2024). https://theintelligentfriend.substack.com/p/generative-ai-makes-you-more-creative.

Vogel, Jeffrey A. "The Speed of Sloth: Reconsidering the Sin of Acedia." *Pro Ecclesia* 18, no.1 (2009): 50-68.

Vogels, Emily A. and Colleen McClain. "Key Findings about Online Dating in the U.S." Pew Research Center (February 2, 2023). https://www.pewresearch.org/short-reads/2023/02/02/key-findings-about-online-dating-in-the-u-s/.

Waldstein, Michael. "Introduction." In *Man and Woman He Created Them: A Theology of the Body*. By John Paul II. Translated by Michael Waldstein. Boston, MA: Pauline Books & Media, 2006: 36-44.

Wang, Hao. "Algorithmic Colonization of Love: The Ethical Challenges of Dating App Algorithms in the Age of AI." *Techné* 27, no.2 (2023): 260-280. https://doi.org/10.5840/techne202381181.

Wang, Hao. "Transparency as Manipulation? Uncovering the Disciplinary Power of Algorithmic Transparency." *Philosophy & Technology* 35, no.3 (2022): 69. https://doi.org/10.1007/s13347-022-00564-w.

Watson, David. "The Rhetoric and Reality of Anthropomorphism in Artificial Intelligence." *Minds and Machines* 29 (2019): 417-440. https://doi.org/10.1007/s11023-019-09506-6.

Webb, Taylor, Keith J. Holyoak, and Hongjing Lu. "Emergent Analogical Reasoning in Large Language Models." *Nature Human Behavior* 7 (2023): 1526-1541. https://doi.org/10.1038/s41562-023-01659-w.

Weber, Max. *Ancient Judaism*. Edited and translated by Hans H. Gerth and Don Martindale. New York: Free Press, 1952.

Weber, Max. *Economy and Society*. Edited by Guenther Roth and Claus Wittich. Berkeley: University of California Press, 1978.

Weber, Max. "Science as a Vocation." In *From Max Weber: Essays in Sociology*. Edited by H.H. Gerth and C. Wright Mills. Oxford: Oxford University Press, 1946.

Weber, Max. *The Protestant Ethic and the Spirit of Capitalism*. Translated by Talcott Parsons. New York: Scribner's, 1958.

Weber, Max. *The Religion of China*. Edited and translated by Hans H. Gerth. New York: Free Press, 1951.

Weber, Max. "The Social Psychology of the World Religions" in *From Max Weber: Essays in Sociology*. Edited and translated by Hans H. Gerth and C. Wright Mills. New York: Oxford University Press, 1958. 267-301.

Wehus, Walter N. "Study Reveals AI-Generated Images Depict Idealized Youth." *Tech Xplore* (October 16, 2024). https://techxplore.com/news/2024-10-

reveals-ai-generated-images-depict.html.

Wei, Jerry, Da Huang, Yifeng Lu, Denny Zhou, Quoc V. Le. "Simple Synthetic Data Reduces Sycophancy in Large Language Models" (2023). https://arxiv.org/abs/2308.03958.

Weizenbaum, Joseph. *Computer Power and Human Reason: From Judgment to Calculation.* Oxford: W. H. Freeman and Co., 1976.

Wenzel, Siegfried. *The Sin of Sloth: Acedia in Medieval Thought and Literature.* Chapel Hill, North Carolina: The University of North Carolina Press, 1967.

Wieringa, Marieke S., Barbara C. N. Müller, Gijsbert Bijlstra, and Tibor Bosse. "Robots Are Both Anthropomorphized and Dehumanized When Harmed Intentionally." *Communications Psychology* 2, no. 72 (2024). https://doi.org/10.1038/s44271-024-00116-2.

Wilson-Hartgrove, Jonathan. *The Wisdom of Stability: Rooting Faith in a Mobile Culture.* Brewster, Mass: Paraclete Press, 2010.

Winn, Zach. "AI Agents Mimic Scientific Collaboration to Generate Evidence-Driven Hypotheses." *Tech Xplore* (December 19, 2024). https://techxplore.com/news/2024-12-ai-agents-mimic-scientific-collaboration.html.

Wojtyła, Karol. *Person and Act and Related Essays*, vol. 1. Translated by Grzegorz Ignatik. Washington, D.C.: The Catholic University of America Press, 2021.

Wong, Matteo. "Science Is Becoming Less Human." *The Atlantic* (December 11, 2023). https://www.theatlantic.com/technology/archive/2023/12/ai-scientific-research/676304.

Wood, David. "Albert Borgmann on Taming Technology: An Interview." *The Christian Century* (August 23, 2003).

Wright, James. "Suspect AI: Vibraimage, Emotion Recognition Technology and Algorithmic Opacity." *Science, Technology, and Society* 28, no.3 (2021): 468-487. https://doi.org/10.1177/09717218211003411.

Wu, Jason Xianghua, Yan Wu, Kay-Yut Chen, and Lei Hua. "Building Socially Intelligent AI Systems: Evidence from the Trust Game Using Artificial Agents with Deep Learning." *Management Science* 69, no.12 (2023): 7236-52. https://doi.org/10.1287/mnsc.2023.4782.

Wu, Philip Fei. "The Privacy Paradox in the Context of Online Social Networking: A Self-Identity Perspective." *Journal of the Association for Information Science and Technology* 70, no.3 (2019): 207-217.

https://doi.org/10.1002/asi.24113.

Wu, T.-Y., and D. J. Atkin. "To Comment or Not to Comment: Examining the Influences of Anonymity and Social Support on One's Willingness to Express in Online News Discussions." *New Media and Society*, 20 (12) (2018): 4512–32.

Xiang, Chloe. "'He Would Still Be Here': Man Dies by Suicide After Talking with AI Chatbot, Widow Says." *Vice* (March 2023). https://www.vice.com/en/article/pkadgm/man-dies-by-suicide-after-talking-with-ai-chatbot-widow-says.

Yang, Yuzhe, Haoran Zhang, Dina Katabi, and Marzyeh Ghassemi. "Change is Hard: A Closer Look at Subpopulation Shift." *Proceedings of Machine Learning Research* 202 (2023): 39584-39622. https://proceedings.mlr.press/v202/yang23s.html.

Zhai, Chunpeng, Santoso Wibowo, and Lily D. Li. "The Effects of Over-Reliance on AI Dialogue Systems on Students' Cognitive Abilities: A Systematic Review." *Smart Learning Environments* 11, no.28 (2024). https://doi.org/10.1186/ s40561-024-00316-7.

Zhang, Shunan, Xiangying Zhao, Tong Zhou, and Jang Hyun Kim. "Do You Have AI Dependency? The Roles of Academic Self-Efficacy, Academic Stress, and Performance Expectations on Problematic AI Usage Behavior." *International Journal of Educational Technology in Higher Education* 21 (2024): 34. https://doi.org/10.1186/s41239-024-00467-0.

Zhang, Yunhao and Renée Gosline. "Human Favoritism, Not AI Aversion: People's Perceptions (and Bias) toward Generative AI, Human Experts, and Human–GAI Collaboration in Persuasive Content Generation." *Judgment and Decision Making* 18 (2023): e41. https://doi.org/10.1017/jdm.2023.37.

Zuboff, Shoshanna. *The Age of Surveillance Capitalism: The Fight for a Human Future at the New Frontier of Power.* New York: Public Affairs, 2019.

Zhou, L., W. Schellaert, F. Martínez-Plumed, et al. "Larger and More Instructable Language Models Become Less Reliable." *Nature* 634 (2024): 61–68. https://doi.org/10.1038/s41586-024-07930-y.

www.ingramcontent.com/pod-product-compliance
Lightning Source LLC
Chambersburg PA
CBHW051416090426
42737CB00014B/2692